T0205866

SpringerBriefs in Mathematical Physics

Volume 19

More information about this series at http://www.springer.com/series/11953

Edouard Brézin · Shinobu Hikami

Random Matrix Theory with an External Source

Edouard Brézin
Laboratoire de Physique Théorique
École Normale Supérieure
Paris
France

Shinobu Hikami
Mathematical and Theoretical Physics Unit
Okinawa Institute of Science and Technology Graduate University
Kunigami-gun, Okinawa
Japan

ISSN 2197-1757 ISSN 2197-1765 (electronic)
SpringerBriefs in Mathematical Physics
ISBN 978-981-10-3315-5 ISBN 978-981-10-3316-2 (eBook)
DOI 10.1007/978-981-10-3316-2

Library of Congress Control Number: 2016961269

Printed on acid-free paper

This Springer imprint is published by Springer Nature
The registered company is Springer Nature Singapore Pte Ltd.
The registered company address is: 152 Beach Road, #22-06/08 Gateway East, Singapore 189721, Singapore

Preface

Random matrix theory (RMT) has a long history, starting with the statistics of nuclear energy levels by Wigner, and it has found applications in wide areas of all sciences from mathematics to biology. The universality of properties derived from random matrices has led to a large number of applications such as network statistics, big data analysis, and biological information. Useful surveys maybe found in [61, 97, 111, 117, 124].

This book deals with Gaussian random matrix models with an external deterministic matrix source. There are also numerous reasons to consider such extensions of standard RMT theory. For instance for a system in the presence of impurities with Hamiltonian $H = H_0 + V$, a Gaussian distribution for the impurity potential V yields a Gaussian distribution for H with a coupling term between H and H_0. However, we have restricted ourselves to a systematic exposition of the main subject that we have studied in earlier publications, namely the geometric properties of surfaces of arbitrary genera deduced from RMT with a nonrandom matrix source.

Indeed for a number of years random matrix theory has been known to be a powerful tool for characterizing geometric properties of surfaces, leading in particular to explicit solution to 2D quantum gravity, and after Kontsevich's resolution of Witten conjectures, to the computation of intersection numbers for curves drawn on Riemann surfaces of given genus with fixed marked points. Many techniques have been used in such problems such as loop equations, Virasoro constraints, and integrable hierarchies. However our work, over a long period, has consisted in showing that simple Gaussian models with an appropriately tuned external source provide an elementary alternative approach to these topics. A systematic exposition of the main tools that underlie this method is the aim of this book. Two properties are basic: (i) the n-point functions are known explicitly for a given arbitrary matrix source. This is true for finite $N \times N$ matrices. (ii) A remarkable duality holds for this probability distribution: the expectation value of the product of K characteristic polynomials over $N \times N$ random matrices is equal to the expectation value of the product of N characteristic polynomials over $K \times K$ random matrices.

Then, an appropriate tuning of the external source generates, in a "double scaling" limit, well-known models such as the Kontsevich Airy matrix model, the Penner model and various generalizations. The duality makes it possible to relate those models to Gaussian models whose correlations functions are known explicitly, providing thereby a simple tool to compute intersection numbers.

Among the mathematical techniques underlying this approach, the Harish Chandra [77], Itzykson-Zuber [80] integral over the unitary group plays a central role. The random matrices there are Hermitian, and the associated random surfaces are orientable. However the general Harish Chandra formula holds for integration over Lie algebras, such as antisymmetric or symplectic matrices for $so(N)$ or $sp(N)$. The corresponding random surfaces are nonorientable. Therefore one can generalize to these Gaussian models in a source the same strategy, namely finding explicit expressions for the correlation functions plus a duality. In the double scaling limit, this provides explicit results for finding the geometric intersection numbers for nonorientable surfaces.

Another extension concerns a supersymmetric duality, whose geometric content is not known to us.

Although the contents mostly consist of a systematic exposition of results scattered through earlier publications, a few new results are presented in the last chapters.

Paris, France Edouard Brézin
Kunigami-gun, Japan Shinobu Hikami

Acknowledgements

We thank our colleagues for discussions about various aspects of matrix models, and referees for comments. S.H. thanks the support of JSPS KAKENHI (C) Grant numbers 25400414 and 16K05491.

About This Book

We consider Gaussian random matrix models in the presence of a deterministic matrix source. In such models, the correlation functions are known exactly for an arbitrary source and for any size of the matrices. The freedom given by the external source allows for various tunings to different classes of universality. Our main interest here is to use this freedom to compute various topological invariants for surfaces, such as the intersection numbers for curves drawn on a surface of given genus with marked points, or the $\mathbf{P}^1$ Gromov–Witten invariants. A remarkable duality for the average of characteristic polynomials is essential for obtaining such topological invariants. We have presented most results in several earlier publications, but here we have attempted to present a unified and concise exposition.

Contents

Chapter 1
Introduction

Random matrix theory is an approach to complex phenomena, such as the energy levels of nuclei, through integrals over $N \times N$ randomly distributed matrix elements. Historically the distribution of the spacings of the eigenvalues was first studied by Wigner [132], and then computed exactly [50, 97]. Dyson [50] conjectured the universality of the spacings distribution (appropriately scaled by the non-universal density of states) which is now well understood, and has also been extended to Gaussian models with a source [18, 22]. Following 't Hooft breakthrough in his study of the $1/N$ expansion of $SU(N)$ gauge theories [125], in which he showed that the large N limit consisted of planar Feynman diagrams and increasing powers in the $1/N$ expansion to diagrams drawn on surfaces of increasing genera, the $1/N$ expansion for random matrices, with non-Gaussian distributions, was introduced and solved in [13]. It was then pointed out that such matrix models could be used as generating functions for counting discretized, for instance triangulated, random surfaces of arbitrary genus [7, 42, 84].

In the early nineties, a study of 2D quantum gravity, the theory of a bosonic string sweeping in its motion a Riemann surface, was then developed on the basis of matrix integrals [15]. Remarkably this discretization of the surface leads, in the continuum limit, to an analytic explicit solution of 2D quantum gravity in terms of integrable hierarchies [14, 55, 69]. Understanding this integrability and universality is a central issue of random matrix theory.

The random Gaussian matrix model with an external source, the central subject of this book, is solvable explicitly and, with appropriately tuned sources, it leads to explicit solutions to a number of geometric problems.

This article consists of 10 chapters. The second deals with Gaussian averages of vertices ; $N \times N$ Hermitian matrices M have invariant vertices of the form $\mathrm{tr} M^k$. The average of products of such vertices gives interesting results depending upon k and N. The Proposition 2.1 gives the generating function for such Gaussian means ; its proof is a consequence of the Theorem 3.3.1, which is one of the main theorems of this book.

E. Brézin and S. Hikami, *Random Matrix Theory with an External Source*,
SpringerBriefs in Mathematical Physics, DOI 10.1007/978-981-10-3316-2_1

In chapter three, an external source matrix A is introduced, which couples to the random matrix M. A key theorem is due to HarishChandra-Itzykson-Zuber, which enables one to integrate over the unitary group. The n-point correlation functions are expressed as a determinant of a kernel (Theorem 3.2.2). This kernel is given in explicit integral form, for finite N and for arbitrary eigenvalues a_i of the external source.

In chapter four, Gaussian averages of the product of K characteristic polynomials are considered. An amazing duality which interchanges K and N is derived (Theorem 4.1.2), and characteristic polynomials for $o(N)$ and $sp(N)$ Lie algebras are considered. There super matrices turn out to be a convenient tool for deriving the correlation functions.

In chapter five, the universality of the random matrix theory is discussed. The universality of the sine kernel, independently of the external source, is shown in Theorem 5.1. The level spacing probability distributions for the sine kernel, for the Airy kernel and generalized Airy kernels, are evaluated. Airy kernel and the generalized Airy kernel are related to the singularities of the Green function. The distribution of the zeros of Riemann's zeta function is known to be related to the level spacing distribution of random matrices [100, 105]. The moments of the characteristic polynomials show the same universal behavior as the average of the moments of Riemann's zeta function and various generalized L-functions.

In chapter six, we discuss the algebraic geometrical study of moduli space of curves, produced by a random matrix model with an external source. As shown by Kontsevich [89] an Airy matrix model, expanded through trivalent ribbon graphs, is a tool for computing the intersection numbers of moduli spaces of curves for given genus.

In chapter seven, curves are generalized to p-spin curves, and the n-point function $U(\sigma_1, ..., \sigma_n)$ are shown to be generating functions of the intersection numbers of the moduli space of p-spin curves for genus g and n marked points.

In chapter eight, the open intersection numbers with a boundary are considered through a Kontsevich-Penner model, a matrix model with a logarithmic term.

In chapter nine, the study of characteristic polynomials of chapter four is further investigated for the Lie algebra $o(N)$ and $sp(N)$ invariant ensembles. Tuning again the external source to a singular $p \to -1$ limit, one recovers a generalized Euler characteristics for non-orientable surfaces.

Finally, in the chapter ten, the matrix model with an external source for the moduli space of curves is extended to the $\mathbf{P}^1$ Gromov-Witten case, providing a computation of its invariants. The Gaussian averages computed in chapter two, are compared, in the large N limit, to the Gromov-Witten invariants of genus zero.

Chapter 2
Gaussian Means

2.1 Integral Representation

In this chapter we are dealing with the standard GUE matrix integrals in the absence of any external source. The Gaussian random matrix theory has a probability distribution $P(M)$

$$P(M) = \frac{1}{Z_0} \exp[-\frac{\lambda}{2} \mathrm{tr} M^2] \tag{2.1}$$

where M is an $N \times N$ Hermitian matrix. The partition function Z_0 is thus given by

$$Z_0 = \int dM \exp[-\frac{\lambda}{2} \mathrm{tr} M^2] \tag{2.2}$$

The integration is performed with the standard $U(N)$ invariant measure over the N^2 matrix elements. The Gaussian average of a product of vertices $\prod \mathrm{tr} M^{k_i}$ is given by

$$\langle \prod_{i=1}^{n} \mathrm{tr} M^{k_i} \rangle = \int dM P(M) \prod_{i=1}^{n} \mathrm{tr} M^{k_i} \tag{2.3}$$

where k_i, $(i = 1, \ldots, n)$ is integer. The Gaussian means may be computed with the help of Wick's theorem, which counts the pairings of the vertices. In Wick's theorem for matrices averages, the size N of the matrices appears in the combinatorics and it is at the origin of the topological properties. The topological dependence on N was first pointed out by t' Hooft [125] for $U(N)$ gauge theories. In the large N limit, planar diagrams dominate. For the non-Gaussian case, planar diagrams are discussed in the approach to two dimensional quantum gravity in [13, 15].

Wick's theorem for the matrix M_{ij} provides the Gaussian means of products of vertices. For instance,

E. Brézin and S. Hikami, *Random Matrix Theory with an External Source*,
SpringerBriefs in Mathematical Physics, DOI 10.1007/978-981-10-3316-2_2

$$\langle \frac{1}{N}\mathrm{tr}M^2\rangle = \langle \frac{1}{N}\sum_{i,j=1}^{N} M_{ij}M_{ji}\rangle = N/\lambda \tag{2.4}$$

$$\langle \frac{1}{N}\mathrm{tr}M^4\rangle = \sum_{i,j,k,l} \frac{1}{N}\langle M_{ij}M_{jk}M_{kl}M_{li}\rangle = \frac{N^2}{\lambda^2}(2+\frac{1}{N^2}) \tag{2.5}$$

Wick's theorem for matrix elements is based upon

$$\langle M_{ij}M_{lk}\rangle = \frac{1}{\lambda}\delta_{i,k}\delta_{j,l} \tag{2.6}$$

where $\delta_{i,k}$ is the Kronecker delta function. In the rest of this section we simply take $\lambda = N$ in the probability measure (2.1). The n-point function, which is a generating function of the above Gaussian means, is given by

$$U(\sigma_1,\ldots,\sigma_n) = \langle \prod_{i=1}^{n} \frac{1}{N}\mathrm{tr}e^{\sigma_i M}\rangle = \frac{1}{N^n}\sum_{k_1,\ldots,k_n} (\prod_{i=1}^{n} \frac{1}{k_i!}\sigma_i^{k_i})\langle \mathrm{tr}M^{k_1}\cdots \mathrm{tr}M^{k_n}\rangle \tag{2.7}$$

This generating function is also the evolution operator of the n-point resolvent G_n, defined as

$$G_n(z_1,\cdots,z_n) = \langle \prod_{a=1}^{n} \frac{1}{N}\mathrm{tr}\frac{1}{z_a - M}\rangle. \tag{2.8}$$

For instance, the average resolvent $G(z)$ is written in terms of the evolution operator as

$$G(z) = \frac{1}{N}\langle \mathrm{tr}\frac{1}{z-M}\rangle = i\int_0^\infty dt e^{-itz}U(\sigma). \tag{2.9}$$

From the definition of (2.7), n-point density correlation function R_n is written as

$$\begin{aligned} R_n(\lambda_1,\ldots,\lambda_n) &= \frac{1}{N^n}\langle \prod_{i=1}^{n} \mathrm{tr}\delta(\lambda_i - M)\rangle \\ &= \int_{-\infty}^{\infty}\cdots\int_{-\infty}^{\infty}\prod_{i=1}^{n}\frac{dt_i}{2\pi}e^{-i\sum_i t_i\lambda_i}U(\sigma_1,\ldots,\sigma_n). \end{aligned} \tag{2.10}$$

We have obtained an exact representation for those generating functions $U(\sigma_1,\ldots,\sigma_n)$, which will be derived in the next section.

Note that $R_n(\lambda_1,\ldots,\lambda_n)$ is obtained also from the probability distribution function $P_N(x_1,\ldots,x_N)$, which becomes after the integration of unitary degree of freedom,

$$P_N(\lambda_1,\ldots,\lambda_N) = C\prod(\lambda_i - \lambda_j)^2 e^{-\frac{1}{2}\sum\lambda_i^2} \tag{2.11}$$

$$R_n(\lambda_1,\ldots,\lambda_n) = \frac{N!}{(N-n)!}\int_{-\infty}^{\infty}\cdots\int_{-\infty}^{\infty} d\lambda_{n+1}\cdots d\lambda_N P_N(\lambda_1,\ldots,\lambda_N) \quad (2.12)$$

This n-point correlation function becomes the determinant of the kernel by the orthogonal polynomial method [97].

$$R_n(\lambda_1,\ldots,\lambda_n) = \det[K_N(\lambda_i,\lambda_j)] \quad (2.13)$$

We will discuss this determinant formula for the case of external source, which orthogonal polynomial method does not work, in Chap. 3 (Theorem 3.2.2).

Proposition 2.1

$$U(\sigma_1,\ldots,\sigma_n) = \frac{1}{N^n}\oint\prod_{i=1}^{n}\frac{du_i}{2\pi i}\prod_{i=1}^{N}(1+\frac{\sigma_i}{u_i})^N e^{\sum u_i\sigma_i/\lambda+\frac{1}{2\lambda}\sum\sigma_i^2}\det\frac{1}{u_i-u_j+\sigma_i} \quad (2.14)$$

where the contours enclose all poles at $u_i = 0$.

This Proposition 2.1 follows from Theorem 3.2.1 in the next section which deals with a Gaussian model with an external source, when the external source vanishes. Remarkably enough introducing a source provides explicit formulae even for a vanishing source, which would be very difficult to derive directly.

For the connected part of the above correlation functions, one simply keeps the longest cycles in the expansion of the determinant which appears in (2.14). The Proposition 2.1 determines explicitly the coefficients of the expansions in powers of σ_i.

In the following, the n-point functions ($n = 1, 2$ and 3) are considered with the help of this formula.

(i) one point function

$$U(\sigma) = \langle\frac{1}{N}\mathrm{tr}e^{\sigma M}\rangle = \frac{e^{\frac{\sigma^2}{2\lambda}}}{N\sigma}\oint\frac{du}{2i\pi}(1+\frac{\sigma}{u})^N e^{u\sigma/\lambda} \quad (2.15)$$

In terms of the density of eigenvalues

$$\rho(x) = \langle\frac{1}{N}\mathrm{tr}\delta(x-M)\rangle \quad (2.16)$$

one has by definition of (σ) in (2.15),

$$U(\sigma) = \int_{-\infty}^{\infty} dx\rho(x)e^{itx} \quad (2.17)$$

where $\sigma = it$. In the large N limit, by the shift of $u \to Nu$ and putting $\lambda = N$, $U(\sigma)$ becomes

$$\lim_{N\to\infty} U(\sigma) = \frac{1}{\sigma}\oint \frac{du}{2i\pi} e^{\sigma(u+\frac{1}{U})} = \sum_{k=0}^{\infty} \frac{1}{k!(k+1)!}\sigma^{2k} = -\frac{1}{i\sigma} J_1(-2i\sigma), \quad (2.18)$$

where $J_1(x)$ is Bessel function of order one. Its Fourier transform provides thus the density of eigenvalues

$$\rho(x) = \int_{-\infty}^{\infty} \frac{dt}{2\pi} e^{-itx} U(it) = \int_{-\infty}^{\infty} \frac{dt}{2\pi}\frac{1}{t} J_1(2t) e^{-ixt} = \frac{1}{\pi}\sqrt{1-\frac{x^2}{4}}, \quad (|x| \leq 2) \tag{2.19}$$

One recovers for the density of state $\rho(x)$, the well known Wigner's semi-circle.

The finite N expression for $U(\sigma)$ is for $\lambda = N$, through a binomial expansion,

$$\begin{aligned} U(\sigma) &= \frac{1}{\sigma}\oint \frac{du}{2i\pi} \sum_{k=0}^{\infty} \frac{N!}{k!(N-k)!u^k} \frac{1}{N^k}\sigma^k \sum_m \frac{1}{m!}\sigma^m u^m e^{\frac{\sigma^2}{2N}} \\ &= e^{\frac{\sigma^2}{2N}} \sum_{k=0}^{\infty} \frac{N!}{N^{k+1}(N-k-1)!k!(k+1)!}\sigma^{2k} \\ &= \sum_{k=0}^{\infty}\sum_{l=0}^{k} \frac{\sigma^{2k}}{2^{k-l}N^k} \frac{(N-1)!}{(N-l-1)!} \frac{1}{l!(l+1)!(k-l)!} \end{aligned} \tag{2.20}$$

This expression, provides an explicit finite N result for the Gaussian average from (2.15) :

$$\begin{aligned} \frac{1}{N}\langle \mathrm{tr} M^{2k}\rangle &= (2k)! \sum_{l=0}^{k} \frac{1}{2^{k-l}N^k} \frac{(N-1)!}{(N-l-1)!} \frac{1}{l!(l+1)!(k-l)!} \\ &= (2k)! \sum_{l=0}^{k} (\prod_{j=1}^{l} (1-\frac{j}{N})) \frac{1}{N} \frac{1}{(2N)^{k-l}} \frac{1}{l!(l+1)!(k-l)!} \end{aligned} \tag{2.21}$$

(Alternatively one may compute the resolvent

$$G(x) = \frac{1}{i}\int_0^{\infty} dt e^{itx} U(it) = -\frac{1}{x}\int_0^{\infty} \frac{dt}{t} e^{it} \oint \frac{du}{2i\pi}(1+\frac{it}{Nu})^N e^{itu/x - t^2/2Nx^2}$$

with $\lambda = N$ in $U(\sigma)$ and expand it in powers of $1/x$.)

The above equations provide also the $1/N$ expansions for $U(\sigma)$ and for the average vertices. The leading large N limit is given by (2.18) and beyond it one finds easily [28]:

$$\prod_{j=1}^{l}(1-\frac{j}{N}) = \exp[\sum_{j=1}^{l}\log(1-\frac{j}{N})]$$
$$= 1 - \frac{l(l+1)}{2N} + \frac{l(l+1)(l-1)(3l+2)}{24N^2} + \cdots \tag{2.22}$$

$$\frac{1}{N}\langle \mathrm{tr}M^{2k}\rangle = \frac{(2k)!}{k!(k+1)!}\Big(1 + \frac{k(k-1)(k+1)}{12N^2} + \frac{k(k+1)(k-1)(k-2)(k-3)(5k-2)}{1440N^4}\Big) + O(\frac{1}{N^6}) \tag{2.23}$$

Okounkov [106, 107] have shown that the Kontsevich intersection numbers [89], may be obtained by taking a simultaneous large N and large k limit of Gaussian averages. From (2.23), the limit for large N, large k, and finite k/N, is

$$\frac{1}{N}\langle \mathrm{tr}M^{2k}\rangle = \frac{(2k)!}{k!(k+1)!}\Big(1 + \frac{1}{12}\frac{k^3}{N^2} + \frac{5}{1440}\frac{k^6}{N^4} + \cdots\Big) \tag{2.24}$$

In the above equation, the coefficients of $\frac{k^{3g}}{N^{2g}}$, namely $\frac{1}{(12)^g g!}$, are the intersection numbers of the moduli space of curves with one marked point.

$$\langle \tau_{3g-2}\rangle_g = \frac{1}{(12)^g g! 2^g} \tag{2.25}$$

In our previous work, we have used the exact integral representation valid for finite N of those vertex correlation functions, and obtained explicitly the scaling region for large k_i and large N by a simple saddle-point [28]. This led to a practical way to compute intersection numbers from a pure Gaussian model, much simpler than with the Kontsevich's Airy matrix model. The intersection numbers derived by several different methods will be discussed in Chap. 6.

(ii) two point function

The connected part is given by the shifts $u_i \to Nu_i$ and $\lambda \to N$ in (2.14),

$$U_c(\sigma_1,\sigma_2) = -\frac{1}{N^2}\oint \frac{du_1 du_2}{(2i\pi)^2}\frac{(1+\frac{\sigma_1}{Nu_1})^N(1+\frac{\sigma_2}{Nu_2})^N}{(u_1-u_2+\frac{\sigma_1}{N})(u_2-u_1+\frac{\sigma_2}{N})} e^{\sigma_1 u_1 + \sigma_2 u_2 + \frac{1}{2N}(\sigma_1^2+\sigma_2^2)} \tag{2.26}$$

Expanding the denominator as

$$\frac{1}{u_2}\sum_{l=0}^{\infty}(\frac{u_1}{u_2})^l(1+\frac{\sigma_1}{Nu_1})^l\sum_{m=0}^{\infty}\frac{u_1^m}{(1+\frac{\sigma_2}{Nu_2})^{m+1}u_2^{m+1}} \tag{2.27}$$

the two point function follows from a residue calculation,

$$
\begin{aligned}
U_c(\sigma_1,\sigma_2) &= \frac{1}{N^2} e^{\frac{1}{2N}(\sigma_1^2+\sigma_2^2)} \sum_{k,t,l,m=0}^{\infty} \frac{\sigma_1^{2k+l+m+1}\sigma_2^{2t+l+m+1}}{k!(k+l+m+1)!t!(t+m+l+1)!} \\
&\times \frac{(N+l)!\,(N-m-1)!}{(N-m-k-1)!(N-m-t-1)!N^{k+t+l+m+1}} \\
&= \frac{1}{N^2} e^{\frac{1}{2N}(\sigma_1^2+\sigma_2^2)} (\sigma_1\sigma_2 + \frac{1}{2}\sigma_1^2\sigma_2^2 + \frac{1}{2}(1-\frac{1}{N})(\sigma_1^3\sigma_2 + \sigma_1\sigma_2^3) + \sigma_1^3\sigma_2^3(\frac{1}{3} - \frac{1}{2N} + \frac{1}{3N^2}) \\
&+ \cdots) \\
&= \frac{1}{N^2}\left(\sigma_1\sigma_2 + \frac{1}{2}\sigma_1^2\sigma_2^2 + \frac{1}{2}(\sigma_1^3\sigma_2 + \sigma_1\sigma_2^3) + (\frac{1}{3} + \frac{1}{12N^2})\sigma_1^3\sigma_2^3 + \cdots\right)
\end{aligned}
\tag{2.28}
$$

Nothing that

$$
U_c(\sigma_1,\sigma_2) = \frac{1}{N^2}\sum_{k_1,k_2} \langle \mathrm{tr}M^{k_1}\mathrm{tr}M^{k_2}\rangle_c \frac{1}{k_1!k_2!}\sigma_1^{k_1}\sigma_2^{k_2} \tag{2.29}
$$

the Gaussian means are obtained,

$$
\begin{aligned}
&\frac{1}{N^2}\langle \mathrm{tr}M\mathrm{tr}M\rangle_c = \frac{1}{N^2},\quad \frac{1}{N^2}\langle \mathrm{tr}M^2\mathrm{tr}\mathrm{M}^2\rangle_c = \frac{2}{N^2}, \\
&\frac{1}{N^2}\langle \mathrm{tr}M\mathrm{tr}M^3\rangle_c = \frac{3}{N^2},\quad \frac{1}{N^2}\langle \mathrm{tr}M^3\mathrm{tr}M^3\rangle_c = \frac{12}{N^2} + \frac{3}{N^4},\ldots
\end{aligned}
\tag{2.30}
$$

(iii) three point function

There are two longest cycles in the determinant (2.14), which contribute to the connected part $U_c(\sigma_1,\sigma_2,\sigma_3)$, which are after the shifts $u_i \to Nu_i$ and $\lambda \to N$,

$$
\begin{aligned}
I &= \frac{1}{(u_1 - u_2 + \frac{\sigma_1}{N})(u_2 - u_3 + \frac{\sigma_2}{N})(u_3 - u_1 + \frac{\sigma_3}{N})} \\
&+ \frac{1}{(u_1 - u_3 + \frac{\sigma_1}{N})(u_3 - u_2 + \frac{\sigma_3}{N})(u_2 - u_1 + \frac{\sigma_2}{N})}
\end{aligned}
\tag{2.31}
$$

Computing the residues in the contour integrals, the three point function reads

$$
\begin{aligned}
U_c(\sigma_1,\sigma_2,\sigma_3) &= -e^{(\sigma_1^2+\sigma_2^2+\sigma_3^2)/2N} \sum_{k,l,m,n_1,n_2,n_3=0}^{\infty} \frac{(N+k)!(N+l)!(N+m)!}{n_1!n_2!n_3!N^{n_1+n_2+n_3+3}} \\
&\times \Bigg(\frac{\sigma_1^{2n_1+k-l}\sigma_2^{2n_2+l-m}\sigma_3^{2n_3+m-k}}{(n_1+k-l)!(n_2+l-m)!(n_3+m-k)!(N+l-n_1)!(N+m-n_2)!(N+k-n_3)!} \\
&+ \frac{\sigma_1^{2n_1+m-l}\sigma_2^{2n_2+k-m}\sigma_3^{2n_3+l-k}}{(n_1+m-l)!(n_2+k-m)!(n_3+l-k)!(N+l-n_1)!(N+m-n_2)!(N+k-n_3)!}\Bigg)
\end{aligned}
\tag{2.32}
$$

2.2 Generating Functions of Gaussian Means

The evaluation of Gaussian means has attracted considerable interests in various fields. A different generating function for the Gaussian means appears for instance in the work of Harer and Zagier [76] and Morozov and Shakirov [102, 103]. Harer and Zagier have found a generating function for $C_{2k} = \frac{1}{N}\langle \mathrm{tr} M^{2k}\rangle$ which reads

$$\sum_{k=0}^{\infty} C_{2k}\frac{N^k x^{2k}}{(2k-1)!!} = \frac{1}{2Nx^2}((\frac{1+x^2}{1-x^2})^N - 1) \tag{2.33}$$

This has the very simple large N limit

$$\lim_{N\to\infty}\sum_{k=0}^{\infty} C_{2k}\frac{x^{2k}}{(2k-1)!!} = \frac{e^{2x^2}-1}{2x^2} \tag{2.34}$$

From (2.33) one finds also

$$\sum_{N,k=0}^{\infty} C_{2k}\frac{x^{2k}\mu^N N^{k+1}}{(2k-1)!!} = \frac{\mu}{1-\mu}\frac{1}{(1-\mu)-(1+\mu)x^2} \tag{2.35}$$

where N is the size of the Hermitian matrix M. These formulae can be derived directly from the above result (2.21). For instance

$$\begin{aligned}\sum_{k=0}^{\infty} C_{2k}\frac{N^k x^{2k}}{(2k-1)!!} &= \int_0^{\infty} dt e^{-t} U(x\sqrt{2Nt}) \\ &= \frac{1}{x}\int_0^{\infty} dt \oint \frac{du}{2i\pi} e^{-t(1-x^2-2Nux)}[(1+\frac{x}{Nu})^N - 1] \\ &= \frac{1}{2Nx^2}((\frac{1+x^2}{1-x^2})^N - 1)\end{aligned} \tag{2.36}$$

Morozov and Shakirov [102] have considered the two point function, for odd powers $C_{2k_1+1,2k_2+1}$, as a generating function of μ,

$$\begin{aligned}&\frac{\mu}{(1-\mu)^{\frac{3}{2}}}\frac{1}{\sqrt{1-\mu+(1+\mu)(x_1^2+x_2^2)}}\arctan(\frac{x_1x_2\sqrt{1-\mu}}{\sqrt{1-\mu+(1+\mu)(x_1^2+x_2^2)}}) \\ &= \sum_{N,k_1,k_2=0}^{\infty} C_{2k_1+1,2k_2+1}\frac{x_1^{2k_1+1}x_2^{2k_2+1}\mu^N}{(2k_1+1)!!(2k_2+1)!!}\end{aligned} \tag{2.37}$$

with $C_{2k_1+1,2k_2+1} = \langle \frac{1}{N}\mathrm{tr}M^{2k_1+1}\frac{1}{N}\mathrm{tr}M^{2k_2+1}\rangle$. These results can also be derived from the generating function (2.15) and (2.26) [103]. The Gaussian means have applications for various problems and further investigated, for instance in [8].

The generating function of Gaussian means is given by the introduction of the parameters t_k coefficients of $\mathrm{tr}M^{k_i}$.

$$Z = \frac{1}{Z_0}\int dM \exp[-\frac{N}{2}\mathrm{tr}M^2 + \sum_{k=1}^{\infty}\frac{t_k}{N}\mathrm{tr}M^k] \tag{2.38}$$

with $Z_0 = \int dMe^{-N/2\,\mathrm{tr}M^2}$. Those Gaussian averages include non-connected parts. Therefore, it is useful to expand the free energy $F = \log Z$. The expansion of F reads (the suffix c indicates the connected parts) :

$$\begin{aligned} F &= \frac{t_1^2}{2N^2}\langle(\mathrm{tr}M)^2\rangle_c + \frac{t_2}{N}\langle\mathrm{tr}M^2\rangle_c + \frac{1}{N^2}t_1t_3\langle\mathrm{tr}M\mathrm{tr}M^3\rangle_c + \frac{t_4}{N}\langle\mathrm{tr}M^4\rangle_c \\ &+ \frac{1}{2N^3}t_1^2t_2\langle(\mathrm{tr}M)^2(\mathrm{tr}M^2)\rangle_c + \frac{t_2^2}{2N^2}\langle(\mathrm{tr}M^2)^2\rangle_c + \cdots \\ &= \frac{1}{2N^2}t_1^2 + t_2 + \frac{3}{N^2}t_1t_3 + t_4(2 + \frac{1}{N^2}) + \frac{1}{N^4}t_1^2t_2 + \frac{1}{N^2}t_2^2 + \cdots. \end{aligned} \tag{2.39}$$

This expansion may be generated with the help of the Virasoro differential operators. They are defined through the identites

$$\int dM\frac{\partial}{\partial M_{ab}}\Big((M^n)_{cd}\exp[-\frac{N}{2}\mathrm{tr}M^2 + \sum_{i=1}^{\infty}\frac{t_i}{N}\mathrm{tr}M^i]\Big) = 0 \tag{2.40}$$

Let us start with the simple $n = 0$ case, the so-called string equation, i.e.

$$\int dM\frac{\partial}{\partial M_{ab}}\Big(\exp[-\frac{N}{2}\mathrm{tr}M^2 + \sum_{k=1}^{\infty}\frac{t_k}{N}\mathrm{tr}M^k]\Big) = 0 \tag{2.41}$$

which gives

$$\langle -NM_{ba} + \frac{1}{N}t_1\delta_{ab} + \frac{1}{N}\sum_{k=2} kt_k(M^{k-1})_{ba}\rangle = 0 \tag{2.42}$$

and after summing on $a = b = 1, \cdots, N$

$$L_{-1}Z = 0 \tag{2.43}$$

with

$$L_{-1} = -N^2\partial_1 + t_1 + \sum(k+1)t_{k+1}\partial_k \tag{2.44}$$

in which ∂_k stands for $\frac{\partial}{\partial t_k}$. Similarly for $n = 1$ the identity(2.40) gives

$$\langle \delta_{ac}\delta_{bd} - NM_{cd}M_{ba} + \frac{1}{N}M_{cd}\sum kt_k(M^{k-1})_{ba}\rangle = 0 \tag{2.45}$$

and, after summing over $b = d$ and $a = c$, one obtains the 'dilaton' equation

$$L_0 Z = 0 \tag{2.46}$$

with

$$L_0 = -N^2\partial_2 + N^2 + \sum_{k=1} kt_k\partial_k \tag{2.47}$$

The same identity for general n provides the full algebra $L_n Z = 0$ for $n \geq -1$ with, for $n \geq 1$,

$$L_n = -N^2\partial_{n+2} + 2N^2\partial_n + N^2\sum_1^{n-1}\partial_p\partial_{n-p} + \sum_1 kt_k\partial_{n+k} \tag{2.48}$$

in which L_1 and L_2 are given by

$$\begin{aligned} L_1 &= -N^2\partial_3 + 2N^2\partial_1 + \sum kt_k\partial_{k+1} \\ L_2 &= -N^2\partial_4 + 2N^2\partial_2 + N^2\partial_1^2 + \sum kt_k\partial_{k+2} \end{aligned} \tag{2.49}$$

These differential operators satisfy the zero central charge Virasoro algebra

$$[L_k, L_m] = L_k L_m - L_m L_k = (k - m)L_{k+m} \tag{2.50}$$

Note that the commutation relations may also be used to determine successively all the L_n for $n \geq 2$. Finally one notes that L_{-1}, L_0, L_1 are linear in the derivatives with respect to the parameters t_k and thus act simply also on the free energy. However the non-linearity of the L_n for $n\rangle 1$ gives non-linear constraints on F. Therefore if the Virasoro constraints can be used easily to determine the lower moments of the Gaussian distribution, the integral form of $U(\sigma_1, \ldots, \sigma_s)$ turns out to be a much more efficient method for computing all the n-points moments.

Instead of Virasoro constraints on the partition functions $Z(t_1, \cdots, t_n, \cdots)$ one can also use recursion relations directly on pure Gaussian means, i.e. with the weight $\exp[-N/2\ \mathrm{trM}^2]$. Consider an operator $\mathcal{O}(M) = \mathrm{trM}^{k_1}\cdots\mathrm{trM}^{k_n}$; one can use systematically the identities

$$\int dM\frac{\partial}{\partial M_{ab}}\Big((M^q)_{cd}\mathcal{O}(M)\exp[-N/2\ \mathrm{trM}^2]\Big) = 0 \tag{2.51}$$

For instance for $q = 0$ one obtains

$$N\langle \mathrm{trM}\,\mathrm{trM}^{k_1}\cdots \mathrm{trM}^{k_n}\rangle = \sum_i k_i \langle \mathrm{trM}^{k_1}\cdots \mathrm{trM}^{k_i-1}\cdots \mathrm{trM}^{k_n}\rangle \tag{2.52}$$

(which follows also immediately from Wick's theorem; $\langle\cdots\rangle$ refers to the expectation value with the normalized weight $\frac{1}{Z_0}\exp[-\frac{N}{2}\,\mathrm{tr}M^2]$ as same as (2.3)). For q=1 similarly for the same $\mathcal{O}(M)$

$$N\langle \mathrm{trM}^2\ \mathcal{O}(\mathrm{M})\rangle = (\mathrm{N}^2 + \sum_i k_i)\langle \mathcal{O}(\mathrm{M})\rangle \tag{2.53}$$

and so on.

2.3 Gaussian Means in the Large N Limit

The Gaussian means in the large N limit are easily derived from the expression of $U(\sigma_1,\ldots,\sigma_n)$, by the replacement

$$\lim_{N\to\infty}\prod_{i=1}^{n}(1+\frac{\sigma_i}{Nu_i})^N = \exp\left(\sum_{i=1}^{n}\frac{\sigma_i}{u_i}\right), \tag{2.54}$$

or from the explicit finite N expressions of Gaussian means in the $N\to\infty$ limit. In this limit, the n-point correlations $U(\sigma_1,\ldots,\sigma_n)$ becomes from (2.14) by the shift $u_i\to Nu_i$ and $\lambda\to N$,

$$U(\sigma_1,\ldots,\sigma_n) = \oint\prod_{i=1}^{n}\frac{du_i}{2\pi i}e^{\sum(u_i+\frac{1}{u_i})\sigma_i}\det\frac{1}{Nu_i-Nu_j+\sigma_i} \tag{2.55}$$

From this formula, we obtain the following Gaussian means in the large N limit.

Proposition 2.3
In the large N limit, the Gaussian connected averages become

$$\langle\frac{1}{N}\mathrm{tr}M^{2k}\rangle = \frac{(2k)!}{k!(k+1)!}$$

$$\langle\frac{1}{N}\mathrm{tr}M^{2k_1}\frac{1}{N}\mathrm{tr}M^{2k_2}\rangle_c = \frac{1}{N^2}\frac{(2k_1)!(2k_2)!}{(k_1!)^2(k_2!)^2}\frac{k_1k_2}{k_1+k_2}$$

$$\langle\frac{1}{N}\mathrm{tr}M^{2k_1+1}\frac{1}{N}\mathrm{tr}M^{2k_2+1}\rangle_c = \frac{1}{N^2}\frac{(2k_1+1)!(2k_2+1)!}{(k_1!)^2(k_2!)^2}\frac{1}{k_1+k_2+1}$$

$$\langle\frac{1}{N}\mathrm{tr}M^{2k_1}\frac{1}{N}\mathrm{tr}M^{2k_2}\frac{1}{N}\mathrm{tr}M^{2k_3}\rangle_c = \frac{1}{N^4}\frac{(2k_1)!(2k_2)!(2k_3)!}{(k_1!)^2(k_2!)^2(k_3!)^2}k_1k_2k_3$$

$$\langle\frac{1}{N}\mathrm{tr}M^{2k_1}\frac{1}{N}\mathrm{tr}M^{2k_2+1}\frac{1}{N}\mathrm{tr}M^{2k_3+1}\rangle_c = \frac{1}{N^4}\frac{(2k_1)!(2k_2+1)!(2k_3+1)!}{(k_1!)^2(k_2!)^2(k_3!)^2}k_1$$

$$\langle \frac{1}{N}\mathrm{tr}M^{2k_1}\frac{1}{N}\mathrm{tr}M^{2k_2}\frac{1}{N}\mathrm{tr}M^{2k_3}\frac{1}{N}\mathrm{tr}M^{2k_4}\rangle_c = \frac{1}{N^6}\prod_{i=1}^{4}\frac{(2k_i)!}{(k_i!)^2}k_1k_2k_3k_4 \times(k_1+k_2+k_3+k_4-1)$$

$$\langle \frac{1}{N}\mathrm{tr}M^{2k_1}\frac{1}{N}\mathrm{tr}M^{2k_2}\frac{1}{N}\mathrm{tr}M^{2k_3+1}\frac{1}{N}\mathrm{tr}M^{2k_4+1}\rangle_c = \frac{1}{N^6}\prod_{i=1}^{4}\frac{1}{(k_i!)^2}(2k_1)!(2k_2)! \times(2k_3+1)!(2k_4+1)!k_1k_2(k_1+k_2+k_3+k_4)$$

$$\langle \frac{1}{N}\mathrm{tr}M^{2k_1+1}\frac{1}{N}\mathrm{tr}M^{2k_2+1}\frac{1}{N}\mathrm{tr}M^{2k_3+1}\frac{1}{N}\mathrm{tr}M^{2k_4+1}\rangle_c = \frac{1}{N^6}\prod_{i=1}^{4}\frac{(2k_i+1)!}{(k_i!)^2} \times(k_1+k_2+k_3+k_4) \quad (2.56)$$

It is easy to check that the Virasoro equation of the previous section holds for the above correlation functions in the large N limit. The Gaussian means appear in various applications. Fat graphs, used in topology (and in biology), use the planar character of the large N limit of Gaussian means [8]. The universal character of Chern-Simon invariants also uses Gaussian means [96]. In Chap. 10, the Gromov-Witten invariants of $\mathbf{P}^1$ will be compared with the expressions of Gaussian means of Proposition 2.3.

2.4 Gaussian Means in the Replica Limit $N \to 0$

We now return to the probability distribution (2.1) with a parameter $\lambda \neq N$. In the study of the intersection numbers of the moduli space of curves, we have used a replica limit ($N \to 0$) in [30] and the reason for this replica limit will be explained below. The integral representation (2.14) for $\lambda \neq N$ reads

$$U(\sigma_1,\ldots,\sigma_n) = \frac{1}{N^n}\oint\prod_{i=1}^{n}\frac{du_i}{2\pi i}(1+\frac{\sigma_i}{u_i})^N e^{\sum u_i\sigma_i/\lambda+\frac{1}{2}\sigma_i^2/\lambda}\det\frac{1}{u_i-u_j+\sigma_i}. \quad (2.57)$$

Let us consider first the $N \to 0$ limit of the one-point function

$$U(\sigma) = \frac{1}{N\sigma}e^{\frac{\sigma^2}{2\lambda}}\oint\frac{du}{2i\pi}e^{\sigma u/\lambda}(1+\frac{\sigma}{U})^N \quad (2.58)$$

from which one gets

$$\lim_{N\to 0}U(\sigma) = \frac{1}{\sigma}e^{\frac{\sigma^2}{2\lambda}}\oint\frac{du}{2i\pi}e^{\sigma u/\lambda}\log(1+\frac{\sigma}{U}) \quad (2.59)$$

The contour integral reduces to the integral of the discontinuity of the logarithm leading to

$$\lim_{N\to 0} U(\sigma) = \frac{\sinh(\frac{\sigma^2}{2\lambda})}{(\frac{\sigma^2}{2\lambda})} \tag{2.60}$$

From above formula, in the replica limit $N \to 0$ of $U(\sigma) = \frac{1}{N}\langle \mathrm{tr} e^{\sigma M}\rangle$, one obtain

$$\lim_{N\to 0} \frac{1}{N}\langle \mathrm{tr} M^k\rangle = \frac{4k!}{\lambda^{2k}4^k(2k+1)!}.$$

Proposition 2.4 ([30])

For the n-point functions the replica limit is given by

$$\lim_{N\to 0} U(\sigma_1,\ldots,\sigma_n) = \frac{\lambda}{\chi^2}\prod_{i=1}^{n} 2\sinh\frac{\chi\sigma_i}{2\lambda} \tag{2.61}$$

with $\chi = \sum_{i=1}^n \sigma_i$.

Using this replica result, the intersection numbers of p-spin curves of the moduli space of curves with one marked point have been derived in [30]. In Kontsevich's model for the intersection numbers of curves, a trivalent vertex $\mathrm{tr}M^3$ has been used. With one marked point, a one stroke line around a marked point characterizes the moduli space of a Riemann surface. This one stroke Feynman diagram is obtained by a $N \to 0$ limit (replica limit).

(2.61) is a generating function for the $N = 0$ limit of $\frac{1}{N}\langle \mathrm{tr}M^{p_1}\cdots \mathrm{tr}M^{p_k}\rangle$ by expanding in the σ_i's. Selecting the coefficients of equal powers for every σ_i, for instance of $(\sigma_1\cdots\sigma_k)^3$, we find

$$\lim_{N\to 0} \frac{1}{N}\langle (\mathrm{tr}M^3)^{4g-2}\rangle = \frac{3^{3g-2}2^{-2g}(6g-4)!}{\lambda^{6g-3}}\cdot\frac{(4g-2)!}{g!(3g-2)!} \tag{2.62}$$

all other powers of $\lim_{N\to 0}\frac{1}{N}\langle (\mathrm{tr}M^3)^k\rangle$ vanishes unless $k = 2$ (mod4). This leads to Kontsevich's intersection numbers $\langle\tau_l\rangle$ for one marked point. Indeed, these intersection numbers $\langle\tau_l\rangle$ are expressed for one marked point, from the partition function Z of following Kontsevich Airy matrix model with $\mathrm{tr}M^3$ term in exponent, as will be shown in Chap. 6,

$$\lim_{N\to 0}\frac{1}{N}\log Z = \sum_{m=1}^{\infty}\langle\tau_m\rangle t_m,\quad Z = \frac{1}{Z'}\int dM e^{\frac{i}{3}\mathrm{tr}M^3 - \mathrm{tr}\Lambda M^2} \tag{2.63}$$

with $Z' = \int dM e^{-\mathrm{tr}M^2}$, and $t_m = (-2)^{-(2m+1)/3}\prod_{l=0}^{m-1}(2l+1)(\frac{2}{\lambda})^{2m+1}$, where all eigenvalues of Λ in Kontsevich model are put equally to λ.

This provides

$$\langle \tau_{3g-2} \rangle_g = \frac{1}{(24)^g g!} \tag{2.64}$$

which agrees with (2.25).

From (2.61), the intersection numbers $\langle \tau_n \rangle$ of one marked point for p-spin curve (the algebraic geomerical definition will be given in Chap. 7) are obtained. The partition function for this higher Airy matrix model with p-spin is

$$Z = \frac{1}{Z_0} \int dM \exp[\frac{1}{p+1} \mathrm{tr}(B^{p+1} - \Lambda^{p+1}) - \mathrm{tr})(B - \Lambda)\Lambda^p] \tag{2.65}$$

with normalization constant Z_0 [30] and $F = \sum_{k_{m,j}} \langle \prod_{m,j} \tau_{m,j}^{k_{m,j}} \rangle \prod_{m,j} t_{m,j}^{k_{m,j}} / k_{m,j}!$, where

$$t_{m,j} = (-p)^{\frac{j-p-m(p+2)}{2(p+1)}} \prod_{l=0}^{m-1} (lp + j + 1) \mathrm{tr} \frac{1}{\Lambda^{mp+j+1}}. \tag{2.66}$$

Expanding the partition function of (2.65) and using (2.61), we obtain the intersection numbers of one marked point by this replica method, for instance we obtain $p = 3$ case,

$$\lim_{N \to 0} \frac{1}{N} \langle (\mathrm{tr} M^4)^{2m-1} \rangle = \frac{2^{2m-2}}{\lambda^{4m-2}} (4m-3)! \frac{(2m-1)!}{m!(m-1)!}, \tag{2.67}$$

$$\langle \tau_{(8g-5-j)/3,j} \rangle_g = \frac{1}{(12)^g g!} \frac{\Gamma(\frac{g+1}{3})}{\Gamma(\frac{2-j}{3})} \tag{2.68}$$

A later section is devoted to the intersection numbers of the moduli spaces of curves by the use of the function of $U(\sigma)$, instead of above formula of $N \to 0$ limit (2.61). In Sect. 5, we will show that, by tuning an external source, the n-point functions $U(\sigma_1, \ldots, \sigma_n)$ are indeed generating functions of the intersection numbers of moduli space of curves. Note that the n-point function of (2.14) is valid only for a Gaussian ensemble without external source. For external source, the formula will be given by (3.6) in Theorem 3.2.1.

Chapter 3
External Source

3.1 HarishChandra Itzykson–Zuber Formula

An external, i.e. deterministic, source matrix A is now coupled to the random matrices with weight

$$P_A(M) = \frac{1}{Z_A} \exp\left[-\frac{\lambda}{2}\mathrm{tr}M^2 + \mathrm{tr}MA\right] \tag{3.1}$$

The matrix A is an $N \times N$ Hermitian matrix. The normalization Z_A, up to a trivial constant, is proportional to $\exp[\frac{1}{2\lambda}\mathrm{tr}A^2]$.

The following integral over the unitary group is due Harish Chandra–Itzykson–Zuber. Consider two Hermitian matrices A and B. Then the integral over the unitary group, with the standard Haar measure normalized to one, is given by the following theorem [77, 80].

Theorem 3.1 (Harish Chandra–Itzykson–Zuber)

$$\int dU \exp(\mathrm{tr}AUBU^\dagger) = c_N \frac{\det[\exp(a_i b_j)]}{\Delta(A)\Delta(B)} \tag{3.2}$$

where $\Delta(A)$ is the Vandermonde determinant $\prod_{i(j}(a_i - a_j)$ constructed with the eigenvalues of A, and c_N is a normalization constant $c_N = \prod_{j=1}^{N-1} j!$.

This formula is a remarkable example of "localization", i.e. the fact that the sum over the saddle points of the integrand is exact.

Proof of Theorem 3.1

This HCIZ formula may be derived by considering the Laplace operator [11, 26],

$$L = -\frac{\partial^2}{\partial X_{ij}\partial X_{ji}} \tag{3.3}$$

Its eigenfunctions are plane waves,

E. Brézin and S. Hikami, *Random Matrix Theory with an External Source*,
SpringerBriefs in Mathematical Physics, DOI 10.1007/978-981-10-3316-2_3

$$Le^{i\mathrm{tr}\Lambda X} = (\mathrm{tr}\Lambda^2)e^{i\mathrm{tr}\Lambda X} \tag{3.4}$$

One can construct a unitary invariant eigenfunction of L, for the same eigenfunction $\mathrm{tr}\Lambda^2$, by the superposition

$$I = \int dU e^{i\mathrm{tr}\Lambda U X U^\dagger} \tag{3.5}$$

which is the HCIZ integral. The integral I being unitary invariant, is a function of the eigenvalues x_i of X. The unitary invariant eigenstates are functions of the eigenvalues x_i of the matrix X and they are orthogonal with the measure

$$\langle\psi_1|\psi_2\rangle = \int dx_1 \cdots dx_N \psi_1^*(x_1, \ldots, x_N)\psi_2(x_1, \ldots, x_N)\Delta^2(x_1, \ldots, x_N)$$

Acting on unitary invariant states the operator L reduces to

$$L = -\frac{1}{\Delta(X)}\sum_i \frac{\partial^2}{\partial x_i^2}\Delta(X).$$

Therefore the lowest eigenstates of this quantum mechanical problems, multiplied by the Vandermonde determinant Δ are simply plane waves; the antisymmetry of Δ under permutations gives thus a free-fermion Slater determinant of plane waves. Therefore the unitary invariant eigenstate I is simply equal to $\det e^{i\lambda_a x_b}/\Delta(X)$ up to a constant. The exchange of Λ and X implies that this constant is proportional to $1/\Delta(\Lambda)$. □

3.2 *n*-Point Function with an External Source

Theorem 3.2.1

$$U(\sigma_1, \ldots, \sigma_n) = \frac{1}{N^n}\oint \prod_{i=1}^n \frac{du_i}{2\pi i}\prod_{\alpha=1}^N \left(1 + \frac{\sigma_i}{u_i - a_\alpha}\right) e^{\sum u_i\sigma_i/\lambda + \sigma_i^2/2\lambda} \det\frac{1}{u_i - u_j + \sigma_i} \tag{3.6}$$

with the contours around all a_α ($\alpha = 1, \ldots, N$), which are eigenvalues of the external source A.

(Proposition 2.1 in the previous section is obviously derived from Theorem 3.2.1 when setting all eigenvalues $a_\alpha = 0$).

The n-point correlation function of $R_n(\lambda_1, \ldots, \lambda_n)$ is defined as (2.10), which is the Fourier transform of the n point function $U(\sigma_1, \ldots, \sigma_n)$. Without external source, this n-point correlation function is expressed in terms of a kernel $K_N(\lambda_i, \lambda_j)$ as

$$R_n(\lambda_1,\ldots,\lambda_n) = \det[K_N(\lambda_i,\lambda_j)] \tag{3.7}$$

where $i, j = 1, \ldots, n$. However, this determinant form remains valid also for the external source problem.

Theorem 3.2.2 *When an external source is coupled to Gaussian distribution, the n-point correlation function of (2.10) is expressed by the determinant of a kernel, as in the sourceless case:*

$$R_n(\lambda_1,\ldots,\lambda_n) = \det[K_N(\lambda_i,\lambda_j)] \tag{3.8}$$

Proof of Theorems 3.2.1 **and** 3.2.2

Given the unitary invariance of the measure dM, one can assume, without loss of generality, that the source matrix A is diagonal, with eigenvalues $(a_1,\ldots,a_N)$.

1. Let us start with the case $n = 1$, i.e.

$$U(\sigma) = \frac{1}{N}\langle \mathrm{tr}\, e^{\sigma M}\rangle \tag{3.9}$$

where $\sigma = it$ and the average is taken with P_A in (3.1), (note that we here normalize $U(\sigma)$ as $U(0) = 1$).

Lemma

$$\int_{-\infty}^{\infty}\prod_{i=1}^{N}\left(\sqrt{\frac{\lambda}{2\pi}}dx_i\right)e^{\sum_{k=1}^{N}\left(-\frac{\lambda}{2}x_k^2+a_kx_k\right)}\Delta(x_1,\ldots,x_N) = Ce^{\frac{1}{2\lambda}\sum_{i=1}^{N}a_i^2}\Delta(a_1,\ldots,a_n) \tag{3.10}$$

in which $\Delta(x_i)$ denotes the Vandermonde determinant $\Delta(x_i) = \prod_{i<j}(x_i - x_j)$ and $C = \lambda^{-N(N-1)/2}$. The proof is easy: after a shift $x_i \to x_i + a_i/\lambda$ which produces the Gaussian factor, one obtains an integral whose result is an antisymmetric polynomial of degree N in the a_i's and thus, up to a constant, it yields $\Delta(a_1\ldots a_N)$. Now we return to (3.9). The change to the variables x_i and Ω, by diagonalizing $M = \Omega X\Omega^\dagger$ yields the Jacobian $\Delta^2(x_1,\ldots x_N)$; the unitary matrix Ω appears only in the exponent as $\mathrm{tr}(\Omega X\Omega^\dagger A)$. The integral over the unitary group is then given by (3.2)

$$U(\sigma) = \frac{1}{Z_A\Delta(A)}\int dx_1\cdots dx_N\Delta(X)(\det e^{a_ix_j})e^{-\sum_j\frac{\lambda}{2}x_j^2}\sum_{k=1}^{N}e^{itx_k} \tag{3.11}$$

The antisymmetry in the N integration variables x_i implies that the $N!$ terms of the determinant give equal contributions and thus up to a factor (the normalization will be given later)

$$U(\sigma) = \frac{1}{Z_A \Delta(A)} \int dx_1 \cdots dx_N \Delta(X) e^{\sum_j (a_j x_j - \frac{\lambda}{2} x_j^2)} \sum_{k=1}^{N} e^{itx_k} \tag{3.12}$$

Defining $\tilde{a}_j^{(k)} = a_j + it\delta_{jk}$ and with lemma (3.10) and the normalization of the probability distribution we obtain

$$\begin{aligned} U(\sigma) &= \frac{1}{N} \sum_{k=1}^{N} e^{\frac{1}{2\lambda} \sum_j [(\tilde{a}_j^{(k)})^2 - a_j^2]} \frac{\Delta(\tilde{A}^{(k)})}{\Delta(A)} \\ &= \frac{1}{N} e^{-t^2/2\lambda} \sum_{k=1}^{N} e^{ita_k/\lambda} \prod_{l \neq k}^{N} \frac{a_k - a_l + it}{a_k - a_l} \end{aligned} \tag{3.13}$$

The r.h.s. may be conveniently written as a contour integral

$$U(\sigma) = \frac{1}{Nit} e^{-t^2/2\lambda} \oint \frac{du}{2i\pi} e^{itu/\lambda} \prod_{k=1}^{N} \left(1 + \frac{it}{u - a_k}\right) \tag{3.14}$$

in which the integral encloses around all the eigenvalues a_i, $(i = 1, \ldots, N)$ of A. The normalization implied by the definition (3.9), i.e. $U(0) = 1$, is manifestly correct, since $\prod_{k=1}^{N}(1 + \frac{it}{u-a_k}) = 1 + it \sum_1^N \frac{1}{u-a_k} + O(t^2)$.

2. The case of the two-point function is similar.

Let us follow the same line for

$$U(\sigma_1, \sigma_2) = \left\langle \frac{1}{N} \mathrm{tr} e^{it_1 M} \frac{1}{N} \mathrm{tr} e^{it_2 M} \right\rangle \tag{3.15}$$

where $\sigma_1 = it_1$ and $\sigma_2 = it_2$. After using Harish Chandra–Itzykson–Zuber formula, it becomes

$$U(\sigma_1, \sigma_2) = \frac{1}{N^2} \sum_{\alpha_1, \alpha_2 = 1}^{N} \int \prod_{i=1}^{N} dx_i \frac{\Delta(x)}{\Delta(A)} e^{\sum_i (-\frac{\lambda}{2} x_i^2 + a_i x_i) + it_1 x_{\alpha_1} + it_2 x_{\alpha_2}} \tag{3.16}$$

where x_i is an eigenvalue of M and a_i of A. Again one can integrate over the x_i with the lemma (3.10)

$$\begin{aligned} U(\sigma_1, \sigma_2) = \frac{1}{N^2} \sum_{\alpha_1, \alpha_2 = 1}^{N} \frac{\prod_{i<j}(a_i - a_j + it_1(\delta_{i,\alpha_1} - \delta_{j,\alpha_1}) + it_2(\delta_{i,\alpha_2} - \delta_{j,\alpha_2})]}{\prod_{i<j}(a_i - a_j)} \\ \times e^{\frac{1}{\lambda}(it_1 a_{\alpha_1} + it_2 a_{\alpha_2} - \frac{1}{2}t_1^2 - \frac{1}{2}t_2^2 - t_1 t_2 \delta_{\alpha_1, \alpha_2})} \end{aligned} \tag{3.17}$$

Dividing the sum into two terms; $\alpha_1 = \alpha_2$ and $\alpha_1 \neq \alpha_2$ one obtains,

$$U(\sigma_1,\sigma_2)=\frac{1}{N^2}\sum_{\alpha_1}\prod_{i(j}\frac{a_i-a_j+i(t_1+t_2)(\delta_{i,\alpha_1}-\delta_{j,\alpha_1})}{a_i-a_j}e^{\frac{i}{\lambda}(t_1+t_2)a_{\alpha_1}-\frac{1}{2\lambda}(t_1+t_2)^2}$$
$$+\frac{1}{N^2}\sum_{\alpha_1\neq\alpha_2}\frac{a_{\alpha_1}-a_{\alpha_2}+i(t_1-t_2)}{a_{\alpha_1}-a_{\alpha_2}}\prod_{\gamma\neq\alpha_1,\alpha_2}\frac{(a_{\alpha_1}-a_\gamma+it_1)(a_{\alpha_2}-a_\gamma+it_2)}{(a_{\alpha_1}-a_\gamma)(a_{\alpha_2}-a_\gamma)}$$
$$\times\exp\frac{1}{\lambda}\left[it_1a_{\alpha_1}+it_2a_{\alpha_2}-\frac{1}{2}(t_1^2+t_2^2)\right] \tag{3.18}$$

Since $R_2(\lambda_1,\lambda_2)$ defined in (2.10) is given by a Fourier transform

$$R_2(\lambda_1,\lambda_2)=\int\frac{dt_1dt_2}{(2\pi)^2}e^{-it_1\lambda_1-it_2\lambda_2}U(\sigma_1,\sigma_2) \tag{3.19}$$

the first sum of (3.18), ($\alpha_1=\alpha_2$ case), becomes δ function, and can be neglected for $R_2(\lambda_1,\lambda_2)$ for $\lambda_1\neq\lambda_2$. The first term of (3.18) is $U(\sigma_1+\sigma_2)/N$ of (3.14), and Fourier transform of this term becomes [17, 18]

$$\frac{1}{N(2\pi)^2}\int\int dt_1dt_2e^{-it_1\lambda_1-it_2\lambda_2}U(\sigma_1+\sigma_2)=\frac{1}{N}\delta(\lambda_1-\lambda_2)\rho(\lambda_1) \tag{3.20}$$

with

$$\rho(\lambda)=\int_{-\infty}^{+\infty}\frac{dt}{2\pi}e^{-it\lambda}U(\sigma). \tag{3.21}$$

where $\sigma_1=it_1,\sigma_2=it_2$ and $\sigma=it$.

The double sum of the second term, ($\alpha_1\neq\alpha_2$ case), is represented by a double contour integral,

$$U(\sigma_1,\sigma_2)=\frac{1}{N^2t_1t_2}e^{-\frac{1}{2\lambda}(t_1^2+t_2^2)}\oint\frac{dudv}{(2i\pi)^2}e^{i(t_1u+t_2v)/\lambda}\frac{(u-v+it_1-it_2)(u-v)}{(u-v+it_1)(u-v-it_2)}$$
$$\times\prod_{\gamma=1}^{N}\left(1+\frac{it_1}{u-a_\gamma}\right)\left(1+\frac{it_2}{v-a_\gamma}\right) \tag{3.22}$$

Since

$$\frac{(u-v+i(t_1-t_2))(u-v)}{(u-v+it_1)(u-v-it_2)}=1-\frac{t_1t_2}{(u-v+it_1)(u-v-it_2)} \tag{3.23}$$

the above result is a sum of a disconnected and a connected part. The disconnected term, which comes from the first term 1 in r.h.s. of above equation, leads to the product of $K_N(\lambda_1,\lambda_1)K_N(\lambda_2,\lambda_2)$. For the connected term, shifting transform t_1 and t_2, $t_1\to t_1+iu$ and t_2+iv, $R_2(\lambda_1,\lambda_2)$ is expressed as

$$R_2(\lambda_1,\lambda_2)=K_N(\lambda_1,\lambda_1)K_N(\lambda_2,\lambda_2)-K_N(\lambda_1,\lambda_2)K_N(\lambda_2,\lambda_1) \tag{3.24}$$

a 2×2 determinant of the kernel, which is expressed by

$$K_N(\lambda_1, \lambda_2) = \frac{1}{N} \int \frac{dt}{2\pi} \oint \frac{du}{2i\pi} \prod_{\gamma=1}^{N} \left(\frac{a_\gamma - it}{u - a_\gamma} \right) \frac{1}{u - it} e^{-\frac{1}{2\lambda}(u^2+t^2) - it\lambda_1 - u\lambda_2}. \tag{3.25}$$

This completes the proof of Theorems 3.2.1 and 3.2.2 for the two-point function.

The generalization of the two point function $R_2(\lambda_1, \lambda_2)$ to $R_n(\lambda_1, \ldots, \lambda_n)$ is straightforward.

$$R_n(\lambda_1, \ldots, \lambda_n) = \frac{1}{N^n} \int \prod_{i=1}^{n} \frac{dt_i}{2i\pi} e^{-\frac{1}{2\lambda}\sum t_p^2 - i\sum t_p \lambda_p} \oint \prod_{i=1}^{n} \frac{du_i}{2i\pi} e^{i\sum t_p u_p}$$
$$\times \prod_{p=1}^{n} \prod_{\alpha=1}^{N} \left(1 + \frac{it_p}{u_p - a_\alpha} \right) \prod_{p=1}^{n} \frac{1}{t_p} \prod_{p<q} \frac{[u_p - u_q + i(t_p - t_q)](u_p - u_q)}{(u_p - u_q + it_p)(u_p - u_q - it_q)} \tag{3.26}$$

With the shift of variables $t_p \to t_p + iu_p$,

$$R_n = \frac{1}{N^n} \int \prod_{i=1}^{n} \frac{dt_i}{2i\pi} \oint \prod_{i=1}^{n} \frac{du_i}{2i\pi} e^{-\frac{1}{2\lambda}\sum t_p^2 - \frac{1}{2\lambda}\sum u_p^2 + \sum \lambda_p(-it_p + u_p)}$$
$$\times \prod_{p=1}^{n} \frac{1}{t_p + iu_p} \prod_{p=1}^{n} \prod_{\alpha=1}^{N} \left(\frac{-a_\alpha + it_p}{u_p - a_\alpha} \right) \prod_{p<q} \frac{(it_p - it_q)(u_p - u_q)}{(-u_q + it_p)(u_p - it_q)}. \tag{3.27}$$

The Cauchy determinant $(i, j = 1, \ldots, n)$ is

$$\det\left[\frac{1}{a_i - b_j} \right] = (-1)^{\frac{n(n-1)}{2}} \frac{\prod_{i<j}(a_i - a_j)(b_i - b_j)}{\prod_{i,j}(a_i - b_j)} \tag{3.28}$$

and from this formula with $a_p = it_p$ and $b_p = u_p$, R_n becomes

$$R_n = \frac{1}{N^n} \int \prod_{i=1}^{n} \frac{dt_i}{2i\pi} \oint \prod_{i=1}^{n} \frac{du_i}{2i\pi} e^{-\frac{1}{2\lambda}\sum(t_p^2 + u_p^2) - \sum \lambda_p(it_p + u_p)} \det\left[\frac{1}{u_j - it_i} \prod_{\gamma=1}^{N} \frac{a_\gamma - it_i}{u_j - a_\gamma} \right] \tag{3.29}$$

Since this determinant id $\det[K(\lambda_i, \lambda_j)]$, Theorems 3.2.1 and 3.2.2 are proved. □

Proposition 3.2.1

This kernel satisfies the composition rule, well-known in the sourceless case,

$$\int_{-\infty}^{+\infty} d\mu K_N(\lambda, \mu) K_N(\mu, \nu) = K_N(\lambda, \nu) \tag{3.30}$$

Proof of Proposition 3.2.1

From the contour integrals,

$$\int_{-\infty}^{+\infty} d\mu K_N(\lambda,\mu)K_N(\mu,\nu) = \int_{-\infty}^{+\infty} \frac{dt_1 dt_2}{(2\pi)^2} \oint \frac{du_1 du_2}{(2i\pi)^2} \prod_{\gamma=1}^{N} \left(\frac{a_\gamma - it_1}{u_1 - a_\gamma}\right)\left(\frac{a_\gamma - it_2}{u_2 - a_\gamma}\right)$$

$$\times \frac{1}{(u_1 - it_1)(u_2 - it_2)} e^{-\frac{1}{2\lambda}(u_1^2+u_2^2+t_1^2+t_2^2)-(it_1\lambda+it_2\mu+u_1\mu+u_2\nu)} \quad (3.31)$$

integration over μ, after the shift $t_2 \to t_2 + iu_1$, provides a δ function for t_2. The contour integration over u_1 around the pole $u_1 = it_1$ reconstructs $K_N(\lambda,\nu)$. □

Chapter 4
Characteristic Polynomials and Duality

4.1 Characteristic Polynomials

The consideration of expectation values of characteristic polynomials was introduced in [85] for the unitary group. We here consider the Hermitian matrix for characteristic polynomials [24]. These expectation values turn out to be often more convenient than resolvents. Some observables such as

$$\overline{G}(\lambda, \mu) = \left\langle \frac{\det(\lambda - M)}{\det(\mu - M)} \right\rangle \tag{4.1}$$

give back to averaged resolvents $G(\mu)$ in (2.9) since

$$G(\mu) = \frac{1}{N} \left\langle \mathrm{tr} \frac{1}{\mu - M} \right\rangle = \frac{1}{N} \frac{\partial}{\partial \lambda} \overline{G}(\lambda, \mu)|_{\lambda=\mu}$$

In addition they satisfy a remarkable duality which will be used in the following.

Theorem 4.1.1 *For Gaussian ensemble of $N \times N$ Hermitian random matrices M in an external matrix source A, the expectation value of the product of K- characteristic polynomials $\langle \prod_{i=1}^{K} \det(\lambda_i \cdot \mathrm{I} - M) \rangle$ is equal to the expectations value by an average of N characteristic polynomials in an ensemble of Gaussian $K \times K$ Hermitian matrices B, with an external matrix source given by the Λ:*

$$\begin{aligned} F_K(\lambda_1, \ldots, \lambda_K) &= \frac{1}{Z_N} \int dM \prod_{i=1}^{K} \det(\lambda_i \cdot \mathrm{I} - M) e^{-\frac{1}{2}\mathrm{tr}(M-A)^2} \\ &= \frac{1}{Z_K} \int dB \prod_{j=1}^{N} \det(-a_j \cdot \mathrm{I} + iB) e^{-\frac{1}{2}\mathrm{tr}(B+i\Lambda)^2} \end{aligned} \tag{4.2}$$

E. Brézin and S. Hikami, *Random Matrix Theory with an External Source*, SpringerBriefs in Mathematical Physics, DOI 10.1007/978-981-10-3316-2_4

with $Z_N = 2^{N/2}\pi^{N^2/2}$, $Z_K = 2^{K/2}\pi^{K^2/2}$; λ_i *is an eigenvalue of* Λ, *and* a_α *an eigenvalue of* A. *This duality has appeared in [24, 30, 72], and has been used to study a* (p,q) *minimal string theory [95].*

Proof of Theorem 4.1.1
As usual one can take, without loss of generality, A diagonal with eigenvalues $(a_1,\dots,a_N)$. The $N\times k$ complex Grassmannian variables $\overline{\psi}_i^\alpha$, ψ_i^α ($i=1,\dots,N$ and $\alpha=1,\dots,K$), normalized for each i and α by

$$\int d\overline{\psi}d\psi\begin{pmatrix}1\\ \overline{\psi}\psi\end{pmatrix}=\begin{pmatrix}0\\1\end{pmatrix} \tag{4.3}$$

The product of characteristic polynomials is then given by the integral

$$\prod_{i=1}^{K}\det(\lambda_i\cdot I-M)=\int\prod_{i=1}^{N}\prod_{\alpha=1}^{K}d\overline{\psi}_i^\alpha d\psi_i^\alpha\exp\left[\sum_{i,j=1}^{N}\sum_{\alpha=1}^{K}\overline{\psi}_i^\alpha(\lambda_\alpha-M)_{ij}\psi_j^\alpha\right] \tag{4.4}$$

This allows to perform the Gaussian average with the identity

$$\frac{1}{Z_N}\int dMe^{-\frac{1}{2}\mathrm{tr}M^2+\mathrm{tr}MX}=e^{\frac{1}{2}\mathrm{tr}X^2} \tag{4.5}$$

The matrix X is here given by

$$X_{pq}=a_p\delta_{p,q}-\sum_{\alpha=1}^{K}\overline{\psi}_q^\alpha\psi_p^\alpha \tag{4.6}$$

and

$$\mathrm{tr}X^2=\mathrm{tr}A^2-2\sum_{p=1}^{N}\sum_{\alpha=1}^{K}a_p\overline{\psi}_p^\alpha\psi_p^\alpha-\sum_{\alpha,\beta=1}^{K}\gamma_{\alpha,\beta}\gamma_{\beta,\alpha} \tag{4.7}$$

with

$$\gamma_{\alpha,\beta}=\sum_{i=1}^{N}\overline{\psi}_i^\alpha\psi_i^\beta. \tag{4.8}$$

The last term of (4.7) is thus $-\mathrm{tr}\gamma^2$, and it is convenient to use the identity

$$e^{-\frac{1}{2}\mathrm{tr}\gamma^2}=\frac{1}{Z_K}\int d\beta e^{-\frac{1}{2}\mathrm{tr}\beta^2+i\mathrm{tr}\gamma\beta} \tag{4.9}$$

where β is an auxiliary Hermitian $K \times K$ matrix. Then $F_K(\lambda_1, \ldots, \lambda_K)$ is written as

$$F_K(\lambda_1, \ldots, \lambda_K) = \frac{1}{Z_K} \int \prod d\overline{\psi}_i^{\alpha} d\psi_i^{\alpha} \int d\beta e^{i\mathrm{tr}\gamma(\beta - i\Lambda) - \frac{1}{2}\mathrm{tr}\beta^2 - \sum_{i=1}^{N} \sum_{\alpha=1}^{K} a_i \overline{\psi}_i^{\alpha} \psi_i^{\alpha}}, \tag{4.10}$$

where Λ is the $K \times K$ diagonal matrix with eigenvalues $(\lambda_1, \ldots, \lambda_K)$. The integration over the Grassmannian variables leads to

$$F_K(\lambda_1, \ldots, \lambda_K) = \frac{1}{Z_K} \int d\beta e^{-\frac{1}{2}\mathrm{tr}\beta^2} \prod_{j=1}^{N} \det[(\lambda_\mu - a_j)\delta_{\mu,\nu} + i\beta_{\mu,\nu}] \tag{4.11}$$

Putting $\beta = B + i\Lambda$, one obtain the duality formula of Theorem 4.1.1. □

The duality formula in Theorem 4.1.1 is for the Gaussian average of characteristic polynomials. For non-Gaussian distributions, the average of the product of characteristic polynomials may also be expressed as a determinant [101].

Proposition 4.1 [24] *For an ensemble of $N \times N$ Hermitian matrices X, with the probability distribution $P(X)$*

$$P(X) = \frac{1}{Z_N} \exp[-N \mathrm{tr} V(X)], \tag{4.12}$$

the average of K distinct characteristic polynomials,

$$F_K(\lambda_1, \ldots, \lambda_K) = \left\langle \prod_{\alpha=1}^{K} \det(\lambda_\alpha - X) \right\rangle, \tag{4.13}$$

may be expressed as a determinant built with orthogonal polynomials

$$F_K(\lambda_1, \ldots, \lambda_K) = \frac{1}{\Delta(\lambda_1, \ldots, \lambda_K)} \det|p_{M-1+i}(\lambda_j)| \tag{4.14}$$

The polynomials $p_n(x)$ are orthogonal with respect to the measure $d\mu(x) = dx e^{-NV(x)}$ and normalized by

$$p_n(x) = x^n + lower \quad order \tag{4.15}$$

Proof of Proposition 4.1
After integration over the unitary group, the average of the product of characteristic polynomials is expressed as

$$F_K(\lambda_1,\ldots,\lambda_K) = \frac{1}{Z_N}\int \prod_{i=1}^{N} d\mu(x_i)\Delta^2(x_1,\ldots,x_N)\prod_{\alpha=1}^{K}\prod_{i=1}^{N}(\lambda_\alpha - x_i) \quad (4.16)$$

$\Delta(x)$ is the Vandermonde determinant, $\Delta(x)=\prod(x_i - x_j)$ and the measure $d\mu(x) = e^{-NV(x)}dx$. Z_N is a normalization constant

$$Z_N = \int \prod_{i=1}^{N} d\mu(x_i)\Delta^2(x_1,\ldots,x_N) \quad (4.17)$$

The integrand can be written as

$$\Delta(x_1,\ldots,x_N)\prod_{\alpha=1}^{K}\prod_{i=1}^{N}(\lambda_\alpha - x_i) = \frac{\Delta(x_1,\ldots,x_N;\lambda_1,\ldots,\lambda_K)}{\Delta(\lambda_1,\ldots,\lambda_K)} \quad (4.18)$$

and

$$\Delta(x_1,\ldots,x_N) = \det[p_n(x_m)] \quad (4.19)$$

since $p_n(x) = x^n +$ lower degree. In (4.19), n runs from 0 to N-1 and m runs from one to N. The numerator in (4.18) is written as

$$\Delta(x_1,\ldots,x_N;\lambda_1,\ldots,\lambda_K) = \det[p_a(u_b)] \quad (4.20)$$

in which a runs from zero to $N+K-1$, b runs from one to $N+K$, and u_b stands for x_b if $b \le N$, or λ_b for $N\langle b \le N+K$. Choosing the polynomials orthogonal with the measure $d\mu(x)$ as

$$\int d\mu(x)p_n(x)p_m(x) = h_n\delta_{n,m} \quad (4.21)$$

one can integrate over the N eigenvalues,

$$\int \prod_{i=1}^{N} d\mu(x_i)\Delta(x_1,\ldots,x_N;\lambda_1,\ldots,\lambda_K)\Delta(x_1,\ldots,x_N)=N!\left(\prod_{n=0}^{N-1} h_n\right)\det[p_\alpha(\lambda_\beta)] \quad (4.22)$$

in which α runs from N to $N+K-1$ and β runs from 1 to K. The normalization factor Z_N is

$$Z_N = \int \prod_{i=1}^{N} d\mu(x_i)\Delta^2(x_1,\ldots,x_N) = N!\left(\prod_{n=0}^{N-1} h_n\right). \quad (4.23)$$

Then $F_K(\lambda_1,\ldots,\lambda_K)$ is written as

$$F_K(\lambda_1, \ldots, \lambda_K) = \frac{1}{\Delta(\lambda_1, \ldots, \lambda_K)} \det \begin{vmatrix} p_N(\lambda_1) & p_{N+1}(\lambda_1) & \cdots & p_{+K-1}(\lambda_1) \\ p_N(\lambda_1) & p_{N+1}(\lambda_1) & \cdots & p_{+K-1}(\lambda_1) \\ & & \cdots & \\ p_N(\lambda_K) & p_{N+1}(\lambda_K) & \cdots & p_{N+K-1}(\lambda_K) \end{vmatrix} \tag{4.24}$$

□

4.2 Characteristic Polynomial in the Real Symmetric Case

The average of the product of the characteristic polynomial for the real symmetric matrix X is

$$F_k(\lambda_1, \ldots, \lambda_k) = \int dX e^{-\frac{N}{2}\mathrm{tr}X^2} \prod_{i=1}^{k} \det(\lambda_i - X) \tag{4.25}$$

where X is a real symmetric matrix, $X = X^T$.

Proposition 4.2.1 ([25])

$$F_k(\lambda_1, \ldots, \lambda_k) = \int dB dD e^{-N\mathrm{tr}(B^2 + D^\dagger D)} [-\mathrm{Pf} M]^N \tag{4.26}$$

where

$$M = \begin{pmatrix} D & \Lambda - iB^T \\ -(\Lambda - iB) & D^\dagger \end{pmatrix} \tag{4.27}$$

in which B is a $k \times k$ Hermitian matrix and D a $k \times k$ complex antisymmetric matrix. Pf is the pfaffian of the antisymmetric matrix M.

The generalization of the HarishChandra formula, for integration over Lie groups, such as real symmetric and quaternion symmetric matrices, is given by the following formula by the use of zonal polynomial Z_p due to Jack [25, 26, 78, 81].

Proposition 4.2.2

$$I_\beta(X, \Lambda) = \int dg \exp(\mathrm{tr} g X g^{-1} \Lambda) \tag{4.28}$$

where g is one of the Lie groups $O(k)$, $U(k)$, $Sp(k)$, and dg is the Haar measure. Let us use the conventional parameter β, which takes value of 1,2 and 4 for $O(k)$, $U(k)$ and $Sp(k)$ respectively. Then the I_β is expressed by the zonal polynomials Z_p,

$$I_\beta(X, \Lambda) = \sum_{m=0}^{\infty} \frac{1}{m!} \left(\prod_{q=0}^{m-1} \frac{1}{1 + q\alpha} \right) \sum_p \chi_p(1) \frac{Z_p(X) Z_p(\Lambda)}{Z_p(1)} \tag{4.29}$$

where $\alpha = \frac{2}{\beta}$, $\chi_p(1)$ *is the character of the representation associated to a Young tableau p, and* $Z_p(X)$ *is a zonal polynomial [93].* $Z_p(1)$ *is a normalization constant, depending on* N.

This I_β *satisfies the differential equation,*

$$\left[\sum_{i=1}^{N} \frac{\partial^2}{\partial x_i^2} + \beta \sum_{i=1,(i\neq j)}^{N} \frac{1}{x_i - x_j} \frac{\partial}{\partial x_i}\right] I = \varepsilon I \tag{4.30}$$

with $\varepsilon = \sum_{i=1}^{N} \lambda_i^2$. *The x-dependent eigenfunctions of this Schrödinger operator have a scalar product given by the measure*

$$\langle \phi_1 | \phi_2 \rangle = \int \prod_{i=1}^{N} dx_i |\Delta(x_1, \ldots, x_N)|^\beta \phi_1^*(x_1, \ldots, x_N) \phi_2(x_1, \ldots, x_N). \tag{4.31}$$

Changing I *to* ψ

$$\psi(x_1, \ldots, x_N) = |\Delta(x_1, \ldots, x_N)|^{\beta/2} I(x_1, \ldots, x_N) \tag{4.32}$$

one obtains the Calogero-Moser integrable Hamiltonian,

$$\left[\sum_{i=1}^{N} \frac{\partial^2}{\partial x_i^2} - \beta\left(\frac{\beta}{2} - 1\right) \sum_{i\langle j} \frac{1}{(x_i - x_j)^2}\right] \psi = \varepsilon \psi. \tag{4.33}$$

In the case $\beta = 2$, *one recovers immediately the HarishChandra-Itzykson-Zuber formula. However for* $\beta = 1$ *and 4, it is difficult to derive a useful formula for random matrix model with an external source, except for small values of the size* N *[26, 78].*

4.3 Characteristic Polynomials Fo the Lie Algebras $o(N)$ and $sp(N)$

Let us consider real anti-symmetric matrices, which are the generators of the Lie group $O(N)$.

Since HarishChandra's formula works for semi-simple Lie algebras, we can consider matrix models made of the elements of a Lie algebra (such as anti-symmetric matrices for $o(N)$), with an external matrix source [24, 31]. The steps exposed above for the unitary case can be repeated for those Lie algebras. The symplectic case is also of interest. We have shown in [23] that Lie algebra $sp(N)$ appears in the description of a superconductor vortex in the dirty limit, and we have also studied antisymmetric matrices coupled to an external source. In the analysis of L functions of type $o(N)$

and $sp(N)$, the characteristic polynomials for such Lie algebras have been studied in the relation to the universal moments of L functions [24]. Since the case of $sp(N)$ has a similar structure of $o(2N)$, we expose mainly here $o(2N)$ case, and HarishChandra's formula for $sp(N)$ and $o(2N+1)$ will be given in Chap. 9, which play the key role for the evaluation of the intersection numbers of non-oriented surfaces.

HarishChandra's formula for a Lie group G reads

$$\int_G e^{\langle Ad(g)\cdot a|b\rangle} dg = \frac{\sum_{w\in W} (\det w) e^{\langle w\cdot a|b\rangle}}{\Delta(a)\Delta(b)} \tag{4.34}$$

where a, b are elements of Cartan subalgebra of Lie algebra h. The generalization of the Vandermonde in the denominator is defined as follows. For any $H \in h$, $\Delta(H) = \prod_{\alpha\in\Delta_+} \alpha(H)$ where Δ_+ is the collection of positive roots, W is the finite reflection group, called the Weyl (or Coxeter) group. $\det(w) = \pm 1$.

For the integration over $SO(2n)$,

$$\int_{SO(2n)} e^{\mathrm{tr}(gag^{-1}b)} dg = C \frac{\sum_{w\in W} (\det w) e^{2\sum_j w(a_j) b_j}}{\prod_{j\langle k} (a_j^2 - a_k^2)(b_j^2 - b_k^2)} \tag{4.35}$$

where $C = (2N-1)! \prod_{j=1}^{2N-1} (2j-1)!$, and w is element of Weyl group, which consists here of permutations followed by the reflections ($y_i \to \pm y_i; i = 1, \ldots, N$) with an even number of sign changes. The matrix H is written as the direct sum of v's defined as $H = h_1 v \oplus h_2 v \oplus \cdots \oplus h_n v$, with $v = \begin{pmatrix} 0 & 1 \\ -1 & 0 \end{pmatrix}$.

The previous duality (4.2) for characteristic polynomials averages, extends to Lie algebras of the classical groups, such as antisymmetric real matrices for the orthogonal Lie algebra $o(2N)$. It reads

$$\left\langle \prod_{\alpha=1}^{k} \det(\lambda_\alpha \cdot \mathrm{I} - X) \right\rangle_A = \left\langle \prod_{n=1}^{N} \det(a_n \cdot \mathrm{I} - Y) \right\rangle_\Lambda \tag{4.36}$$

where X is a $2N \times 2N$ real antisymmetric matrix ($X^t = -X$) and Y a $2k \times 2k$ real antisymmetric matrix; the eigenvalues of X and Y are thus pure imaginary. The matrix source A is also a $2N \times 2N$ antisymmetric matrix, coupled to X by $\mathrm{tr}XA$. The matrix Λ is $2k \times 2k$ antisymmetric matrix, coupled to Y. We assume, without loss of generality, that A and Λ have the canonical form: $A = a_1 v \oplus \cdots \oplus a_N v$.

$$A = \begin{pmatrix} 0 & a_1 & 0 & 0 & \cdots \\ -a_1 & 0 & 0 & 0 & \cdots \\ 0 & 0 & 0 & a_2 & 0 \\ 0 & 0 & -a_2 & 0 & 0 \\ \cdots & & & & \end{pmatrix}, \tag{4.37}$$

Λ is expressed also as $\Lambda = \lambda_1 v \oplus \cdots \oplus \lambda_k v$. The characteristic polynomial $\det(\lambda \cdot I - X)$ has the $2N$ roots, $(\pm i\lambda_1, \cdots, \pm i\lambda_n)$. The Gaussian averages in (4.36) are defined as $\langle \cdots \rangle_A = \frac{1}{Z_A} \int dX e^{\frac{1}{2}\mathrm{tr}X^2 + \mathrm{tr}XA}$, $\langle \cdots \rangle_\Lambda = \frac{1}{Z_\Lambda} \int dY e^{\frac{1}{2}\mathrm{tr}Y^2 + \mathrm{tr}Y\Lambda}$ in which X is a $2N \times 2N$ real antisymmetric matrix, and Y a $2k \times 2k$ real antisymmetric matrix; the coefficients Z_A and Z_Λ are such that the expectation values are normalized to one. The derivation relies on a representation of the characteristic polynomials in terms of integrals over Grassmann variables, as for the $U(N)$ case, but it is more involved [31].

Here again the HarishChandra formula leads to explicit formulae for the correlation functions. The one-point function for instance is

$$\begin{aligned} U(\sigma) &= \frac{1}{2N}\langle \mathrm{tr} e^{\sigma X} \rangle_A \\ &= \frac{1}{N\sigma} \oint \frac{du}{2\pi i} \prod_{n=1}^{N} \left(\frac{(u + \frac{\sigma}{2})^2 - a_n^2}{u^2 - a_n^2} \right) \frac{u}{u + \frac{\sigma}{4}} e^{u\sigma + \frac{\sigma^2}{4}}, \end{aligned} \tag{4.38}$$

where the contour encircles the poles $u = a_n$. One may repeat the same tuning plus duality strategy in this case, leading to the desired topological numbers for non-orientable surfaces generated by these antisymmetric matrix models,which will be discussed in Chap. 9. The cases of $o(2N+1)$ and $sp(N)$ have similar duality formula as $o(2n)$, and one point function and the intersection numbers for such cases will be derived in Chap. 9.

4.4 Characteristic Polynomials for Supermatrices

This brief introduction to supermatrices and super determinants intends to show that they are convenient tools to express the characteristic polynomials of supermatrices and their averaged resolvent correlation functions G_N in (2.8) [27, 130].

Consider the expectation value of a ratio of characteristic polynomials

$$F_{P,Q}(\lambda_\alpha \cdots \mu_\beta \cdots) = \frac{1}{Z_N} \left\langle \frac{\prod_{\alpha=1}^{P} \det(\lambda_\alpha - M)}{\prod_{\beta=1}^{Q} \det(\mu_\beta - M)} \right\rangle_A \tag{4.39}$$

with

$$\langle \mathcal{O}(M) \rangle_A = \frac{1}{Z_A} \int dM \mathcal{O}(M) e^{-\frac{1}{2}\mathrm{tr}M^2 + \mathrm{tr}MA} \tag{4.40}$$

Such expectation values are interesting; for instance, the average resolvent is given by $P = Q = 1$, after taking derivative with respect to λ and setting $\lambda = \mu$.

Let us recall standard definitions for supermatrices: let

$$X = \begin{pmatrix} a & \alpha \\ \beta & b \end{pmatrix} \tag{4.41}$$

in which the matrix elements of a and b are commuting numbers, those of α and β are anticommuting. Then the supertrace

$$\mathrm{str} X = \mathrm{tr} a - \mathrm{tr} b \tag{4.42}$$

ensures the cyclic invariance. The super determinant is given by

$$\mathrm{sdet} X = \frac{\det a}{\det(b - \beta a^{-1}\alpha)} = \frac{\det(a - \alpha b^{-1}\beta)}{\det b} \tag{4.43}$$

based on the integral

$$\int d\theta d\bar{\theta} dx d\bar{x} \varepsilon^{i\bar{\Phi} X \Phi} = (\mathrm{sdet} X)^{-1} \tag{4.44}$$

where $\Phi = \begin{pmatrix} x \\ \theta \end{pmatrix}$, $\overline{\Phi} = (\bar{x}\ \bar{\theta})$. The formulae are obtained either by integrating first the commuting variables, or the anticommuting variables first. We use the conventions $\overline{\theta_1\theta_2} = \bar{\theta}_1\bar{\theta}_2$ and $\bar{\bar{\theta}} = \theta$.

Finally the usual bosonic formula still holds here, namely

$$\mathrm{str}(\log X) = \log(\mathrm{sdetX}). \tag{4.45}$$

We are now in position to derive the duality formula for (4.39) which we first write in integral form as

$$F_{P,Q}(\lambda_\alpha \cdots \mu_\beta \cdots) = \int \prod_{a=1}^{N} \prod_{\alpha=1}^{P} \prod_{\beta=1}^{Q} d\bar{x}_\alpha^a dx_\alpha^a d\bar{\theta}_\beta^a d\theta_\beta^a$$
$$\langle e^{-\sum_{\alpha=1}^{P} \bar{x}_\alpha^a (\lambda_\alpha - M)_{ab} x_\alpha^b - \sum_{\beta=1}^{Q} \bar{\theta}_\beta^a (\mu_\beta - M)_{ab} \theta_\beta^b} \rangle_A \tag{4.46}$$

or, introducing the $(Q+P) \times (Q+P)$ diagonal matrix Λ made of μ_β and λ_α,

$$F_{P,Q}(\lambda_\alpha \cdots \mu_\beta \cdots) = \int \prod_{a=1}^{N} \prod_{\alpha=1}^{P} \prod_{\beta=1}^{Q} d\bar{x}_\alpha^a dx_\alpha^a d\bar{\theta}_\beta^a d\theta_\beta^a \langle e^{-\bar{\Phi}^a \Lambda \Phi^a + \bar{\Phi}^a M_{ab} \Phi^b} \rangle_A \tag{4.47}$$

Since

$$\langle e^{\mathrm{tr} XM} \rangle_A = e^{\frac{1}{2}\mathrm{tr} X^2 + \mathrm{tr} AX} \tag{4.48}$$

we have

$$X_{ba} = \bar{\Phi}^a \cdot \Phi^b = \sum_{\alpha=1}^{P+Q} \bar{\Phi}^a_\alpha \Phi^b_\alpha. \tag{4.49}$$

Then

$$\mathrm{tr}(AX) = \sum_{n=1}^{N} a_n \sum_{\alpha=1}^{P+Q} \bar{\Phi}^n_\alpha \Phi^n_\alpha \tag{4.50}$$

in which the a_n are the eigenvalues of A,

$$\mathrm{tr}X^2 = \sum_{a,b=1}^{N} \sum_{\alpha,\beta=1}^{P+Q} \bar{\Phi}^a_\alpha \Phi^b_\alpha \bar{\Phi}^b_\beta \Phi^a_\beta. \tag{4.51}$$

Let us define the matrix Γ, $(Q+P) \times (Q+P)$

$$\Gamma_{\alpha,\beta} = \sum_{a=1}^{N} \overline{\Phi^a_\alpha} \Phi^a_\beta = \begin{pmatrix} \Gamma_1 & \Gamma_2 \\ \Gamma_2^\dagger & \Gamma_3 \end{pmatrix} = \begin{pmatrix} \bar{x} \cdot x & \bar{\theta} \cdot x \\ \bar{x} \cdot \theta & \bar{\theta} \cdot \theta \end{pmatrix} \tag{4.52}$$

This matrix Γ_1 is Hermitian but Γ_3 is anti-Hermitian.

To express $\mathrm{tr}X^2$ in terms of the matrix Γ some commutations are required and one obtains easily

$$\mathrm{tr}X^2 = \sum_{\alpha,\beta} \mathrm{str}(\Gamma^2) \tag{4.53}$$

Therefore

$$F_{P,Q}(\lambda_\alpha \cdots \mu_\beta \cdots) = \int \prod_{a=1}^{N} d\bar{x}^a_\alpha dx^a_\alpha d\bar{\theta}^a_\beta d\theta^a_\beta \\ e^{-\bar{\Phi}^a \Lambda \Phi^a + \sum_{n=1}^{N} a_n \sum_{\alpha=1}^{P+Q} \bar{\Phi}^n_\alpha \Phi^n_\alpha + \frac{1}{2}\mathrm{str}\Gamma^2} \tag{4.54}$$

The SUSY Hubbard-Stratonovich transformation reads

$$\int d\Delta e^{\mathrm{str}(-\frac{1}{2}\Delta^2 + \Delta\Gamma)} = e^{\frac{1}{2}\mathrm{str}\Gamma^2} \tag{4.55}$$

in which Δ is $(P+Q) \times (P+Q)$ and like Γ as far as hermiticity is concerned. Then

$$F_{P,Q}(\lambda_\alpha \cdots \mu_\beta \cdots) = \int d\Delta \int \prod_{a=1}^{N} d\bar{x}^a_\alpha dx^a_\alpha d\bar{\theta}^a_\beta d\theta^a_\beta \\ e^{-\bar{\Phi}^a \Lambda \Phi^a + \sum_{n=1}^{N} a_n \sum_{\alpha=1}^{P+Q} \bar{\Phi}^n_\alpha \Phi^n_\alpha} e^{-\frac{1}{2}\mathrm{str}\Delta^2 + \mathrm{str}\Delta\Gamma} \tag{4.56}$$

One can integrate out on the x's and θ's. The quadratic form in the exponential is

$$-\bar{\Phi}^a \Lambda \Phi^a + \sum_{n=1}^{N} a_n \sum_{\alpha=1}^{P+Q} \bar{\Phi}^n_\alpha \Phi^n_\alpha + \Delta_{\alpha,\beta} \bar{\Phi}^a_\beta \Phi^a_\alpha (-1)^{F_\beta}$$

in which $F_\beta = 0$ for $1 \le \beta \le P$ or $F_\beta = 1$ for $(P+1) \le \beta \le (P+Q)$ The integration then gives

$$\prod_1^N s\det{}^{-1}[(\Lambda_\alpha - a_n)\delta_{\alpha\beta} - \Delta_{\alpha\beta}(-1)^{F_\beta}]$$

Therefore we change $\Delta_{\alpha\beta}(-1)^{F_\beta} \to \tilde{\Delta}_{\alpha\beta}$ and one verifies that

$$\mathrm{str}\Delta^2 = \mathrm{str}\tilde{\Delta}^2. \tag{4.57}$$

Then one ends up with the following proposition,

Proposition 4.4

$$F_{P,Q}(\lambda_\alpha \cdots \mu_\beta \cdots) = \frac{1}{Z_N} \left\langle \frac{\prod_{\alpha=1}^{P} \det(\lambda_\alpha - M)}{\prod_{\beta=1}^{Q} \det(\mu_\beta - M)} \right\rangle_A$$

$$= \int d\Delta e^{-\frac{1}{2}\mathrm{str}\Delta^2} \prod_1^N \mathrm{sdet}^{-1}[(\Lambda_\alpha - a_n)\delta_{\alpha\beta} - \Delta_{\alpha\beta}] \tag{4.58}$$

The above identity (4.58) relates an ordinary integral to a super matrix integration. In this sense it is not a full duality although it can be used for the large N-limit or for a super-generalization of the Kontsevich model. However a full super duality has been derived by Desrosiers and Eynard for expectation values of ratios of super determinants [51] and our identity appears as a simple limiting case [87].

Arbitrary β

An extension of the GUE duality (4.2) to the three classical Gaussian ensembles GOE, GUE, GSE with respectively $\beta = 1, 2, 4$ has been derived by Desrosiers [52], but it exchanges β to $4/\beta$. However the lack of HarishChandra formula for integrating over the orthogonal or symplectic group in terms of $\tau_{ij} = (x_i - x_j)(\lambda_i - \lambda_j)$ $(i, j = 1, \ldots, N)$ for general N [26, 78] does not allow one to compute explicitly the k-point functions and we cannot repeat the steps that we have followed for $\beta = 2$. However we have used supergroup methods to obtain the one and two-point functions [23, 27, 29].

Chapter 5
Universality

5.1 Universal Correlation Functions

The sine kernel was derived for the Gaussian unitary ensemble (GUE) by Dyson [50].

$$K_N(\lambda, \mu) = -\frac{1}{\pi N(\lambda - \mu)} \sin[\pi N(\lambda - \mu)\rho(\lambda)] \tag{5.1}$$

where $\rho(\lambda)$ is the density of states, i.e. the one point function. More generally for arbitrary unitary invariant ensembles the correlation functions are known to have a universal scaling limit as follows, apart from a scale dependence provided by the density of states ρ. For instance the two point connected correlation function of two eigenvalues λ and μ has a universal scaling limit, when the distance $\lambda - \mu$ is measured in terms of the average spacing $1/N\rho$, in other words when $N \to \infty$ and the distance to zero, so that $N(\lambda - \mu)$ is fixed.

This universality holds for unitary invariant measures, but it is also true in the presence of an arbitrary external source matrix A.

Theorem 5.1 *For the GUE with an external source, the sine kernel (5.1) is independent of the external source A.*

Proof of Theorem 5.1 [18] From the integral expression of the kernel $K_N(\lambda, \mu)$ derived in (3.25), the large N limit, $N \to \infty$ but $N(\lambda - \mu)$ fixed, is given by a saddle-point. The density of states of the external source $\rho_0(\lambda)$ is denoted

$$\rho_0(\lambda) = \frac{1}{N} \sum_{i=1}^{N} \delta(\lambda - a_i). \tag{5.2}$$

The unperturbed resolvent is

$$G_0(z) = \frac{1}{N} \mathrm{tr} \frac{1}{z - A} = \int da \frac{\rho_0(a)}{z - a}. \tag{5.3}$$

E. Brézin and S. Hikami, *Random Matrix Theory with an External Source*, SpringerBriefs in Mathematical Physics, DOI 10.1007/978-981-10-3316-2_5

Since from (3.13),

$$U(-\sigma) = -\frac{1}{it}\oint \frac{du}{2i\pi}\prod_{\gamma=1}^{N}\left(1 - \frac{it}{N\left(u-a_\gamma\right)}\right)e^{-itu-\frac{t^2}{2N}} \tag{5.4}$$

where $\sigma = it$, and in the large N limit, the integrand of (5.4) becomes

$$\prod_{\gamma=1}^{N}\left(1 - \frac{it}{N(u-a_\gamma)}\right) = \exp\sum_{\gamma=1}^{N}\log\left(1 - \frac{it}{N\left(u-a_\gamma\right)}\right)$$
$$\sim \exp\left[-\sum_{\gamma}\frac{it}{N(u-a_\gamma)}\right] = \exp[-itG_0(u)] \tag{5.5}$$

and $t^2/2N$ is negligible. Therefore the average resolvent,

$$G(z) = \langle\frac{1}{N}\mathrm{tr}\frac{1}{\mathrm{z}-\mathrm{M}}\rangle = -\int_0^\infty d\tau \mathrm{U}(-i\tau)\exp(-\tau z)$$

satisfies

$$\frac{\partial G}{\partial z} = \oint \frac{du}{2i\pi}\frac{1}{u+G_0(u)-z} \tag{5.6}$$

The contour circles over the N-poles given by the solution of

$$u + \frac{1}{N}\sum_{\gamma}\frac{1}{u-a_\gamma} = z \tag{5.7}$$

which remain close to the a_γ for z large. In addition there is an $(N+1)$th solution $\hat{u}(z)$, which goes to infinity with z

$$\hat{u}(z) = z - \frac{1}{z} + O\left(\frac{1}{z^2}\right) \tag{5.8}$$

Therefore, instead of summing over the first N roots, one takes the contribution of this external pole $\hat{u}(z)$, plus the pole at infinity. Then

$$\frac{\partial G}{\partial z} = 1 - \frac{1}{1+\frac{dG_0}{d\hat{u}(z)}} = 1 - \frac{d\hat{u}(z)}{dz} \tag{5.9}$$

since $u + G_0(u) = z$. The integration of this equation is

$$G(z) = z - \hat{u}(z) \tag{5.10}$$

This leads to Pastur's self consistent equation [113].

$$G(z) = G_0(z - G(z)) \tag{5.11}$$

By the same argument,

$$\frac{\partial K_N}{\partial \lambda_1} = \frac{1}{\pi} \mathrm{Im} \oint \frac{du}{2i\pi} \frac{1}{u + G_0(u) - \lambda_1 + i\varepsilon} e^{-uy} \tag{5.12}$$

with $y = N(\lambda_1 - \lambda_2)$. The term t^2/N is neglected. Then, from (5.6) and (5.9), with $\hat{u} = \hat{u}(\lambda_1 - i\varepsilon)$, we obtain

$$\frac{\partial K_N}{\partial \lambda_1} = \frac{1}{\pi} \mathrm{Im} \frac{d\hat{u}}{d\lambda_1} e^{-y\hat{u}(\lambda_1 - i\varepsilon)} = -\frac{1}{\pi y} \frac{\partial}{\partial \lambda_1} \mathrm{Im}(e^{-y\hat{u}(\lambda_1 - i\varepsilon)}) \tag{5.13}$$

Since

$$\hat{u}(\lambda_1 - i\varepsilon) = \lambda_1 - \mathrm{Re}G(\lambda_1) - i\pi\rho(\lambda_1) \tag{5.14}$$

$$K_N(\lambda_1, \lambda_2) = -\frac{1}{\pi y} e^{-y(\lambda_1 - \mathrm{Re}G(\lambda_1))} \sin[\pi y \rho(\lambda_1)] \tag{5.15}$$

The phase factor of above expression, which differs from (5.1), is cancelled when combined with the conjugate $K_N(\lambda_2, \lambda_1)$:

$$K_N(\lambda_1, \lambda_2) K_N(\lambda_2, \lambda_1) = |\frac{\sin(\pi N \rho(\lambda_1)(\lambda_1 - \lambda_2))}{\pi N(\lambda_1 - \lambda_2)}|^2 \tag{5.16}$$

□

5.2 Level Spacing Probability Distribution

Let us consider the probability $E(\theta)$ that the interval $\left[-\frac{1}{2}\theta, \frac{1}{2}\theta\right]$ does not contain any of the eigenvalues $(x_1, \ldots, x_N)$

$$E(\theta) = \int_{out} \cdots \int_{out} P_N(x_1, \ldots, x_N) dx_1 \cdots dx_N \tag{5.17}$$

with the probability distribution of the N eigenvalues, $x_1, \ldots., x_N$,

$$P_N(x_1, \ldots., x_N) = C \prod_{i(j} (x_i - x_j)^2 e^{-N \sum_{i=1}^{N} V(x_i)},$$

where C is a normalization constant and $V(x)$ is a polynomial of x. The integral being is performed outside of the interval $\left[-\frac{1}{2}\theta, \frac{1}{2}\theta\right]$ [97],

$$\int_{out} dx = \left(\int_{-\infty}^{\infty} - \int_{-\frac{\theta}{2}}^{\frac{\theta}{2}}\right) dx. \tag{5.18}$$

and the n-point correlation function $R_n(x_1, \ldots, x_n)$ in (2.12) is

$$R_n(x_1, \ldots, x_n) = \frac{N!}{(N-n)!} \int_{-\infty}^{\infty} \cdots \int_{-\infty}^{\infty} dx_{n+1} \cdots dx_N P_N(x_1, \ldots, x_N). \tag{5.19}$$

Then $E(\theta)$ may be expanded as

$$E(\theta) = 1 - N \int_{-\frac{\theta}{2}}^{\frac{\theta}{2}} \rho(x) dx + \frac{N^2}{2!} \int_{-\frac{\theta}{2}}^{\frac{\theta}{2}} \int_{-\frac{\theta}{2}}^{\frac{\theta}{2}} R_2(x, y) dx dy + \cdots . \tag{5.20}$$

The natural scale for the level spacing θ is of order $\frac{1}{N}$. Then the short distance scaling limit is defined by θ goes to zero and N to infinity, with fixed $N\theta$. In this limit

$$N \int_{-\frac{\theta}{2}}^{\frac{\theta}{2}} \rho(x) = N\theta\rho(0) + O\left(\frac{1}{N}\right). \tag{5.21}$$

Therefore one replaces θ by the scaling variable s

$$s = N\theta\rho(0) \tag{5.22}$$

In this scaling limit, $E(s)$ is expressed as

$$E(s) = \sum_{n=0}^{\infty} \frac{(-1)^n}{n!} \int_{-\frac{s}{2}}^{\frac{s}{2}} \cdots \int_{-\frac{s}{2}}^{\frac{s}{2}} dx_1 \cdots dx_n \det[\overline{K}(x_i, x_j)] \tag{5.23}$$

where $\overline{K}(x, y)$ is

$$\overline{K}(x, y) = \frac{\sin[\pi(x-y)]}{\pi(x-y)} \tag{5.24}$$

This leads for small s to the expansion

$$E(s) = 1 - s + \frac{\pi^2}{36} s^4 - \frac{\pi^4}{675} s^6 + O(s^7) \tag{5.25}$$

Let us now introduce the eigenvalues $\lambda_i(s)$ of the kernel $\tilde{K}$: Then

$$\int_{-\frac{s}{2}}^{\frac{s}{2}} \overline{K}(x,y)\psi_i(y)dy = \lambda_i \psi_i(x) \tag{5.26}$$

$$E(s) = \prod_{i=1}^{\infty}(1-\lambda_i) = \det[1-\overline{K}] \tag{5.27}$$

The level spacing probability distribution $p(s)$ is obtained by

$$p(s) = \frac{d^2}{ds^2}E(s). \tag{5.28}$$

A closed equation for $E(s)$ for the sine kernel was obtained by Jimbo et. al. [83]. Since, as shown above, a generalized kernel governs also the case of an external source, one can generalize $E(s)$ and $p(s)$ to the external source problem. Furthermore, as will be discussed in Sect. 5.3, an Airy kernel governs the edge of the spectrum. For that problem Tracy and Widom have obtained $p(s)$ using a Fredholm determinant method [126]. Other kernels, such as the one which governs a gap closing distribution, are of interest and the same technique can be used to handle those generalized kernels.

The Fredholm determinant relative to an interval $[a, b]$, is given by the expansion of the determinant (5.27)

$$E(a,b) = \sum_{n=0}^{\infty} \frac{(-1)^n}{n!} \int_a^b \cdots \int_a^b \prod_{k=1}^{n} dx_k \det[\overline{K}(x_i, x_j)] \tag{5.29}$$

where $i, j = 1, \ldots, n$. Extending the analysis of Tracy and Widom [126], a Hamiltonian system may be derived, which leads to coupled nonlinear differential equations. We now follow the presentation of [22].

Considered a kernel, relative to an interval (a, b), of the form

$$K(x,y) = \frac{\phi(x)\phi'(y) - \phi(y)\phi'(x)}{x-y}. \tag{5.30}$$

For the usual sine kernel $\phi(x) = \sin x$, and $\phi''(x) = -\phi(x)$. It is the structure (5.30) which generalizes to edge problems and source problems. The operator $\hat{K}$ is introduced as

$$[X, \hat{K}] = |\phi\rangle\langle\phi'| - |\phi'\rangle\langle\phi| \tag{5.31}$$

in which X is a position operator. It is convenient to define $q(x)$ and $p(x)$ as

$$q(x) = \langle x|\frac{1}{1-\hat{K}}|\phi\rangle, \quad p(x) = \langle\phi'|\frac{1}{1-\hat{K}}|x\rangle \tag{5.32}$$

and the Fredholm resolvent $\tilde{K}$ by

$$\tilde{K} = \frac{\hat{K}}{1-\hat{K}}. \tag{5.33}$$

Note the simple relation,

$$(x-y)\tilde{K} = \langle x|[X,\tilde{K}]|y\rangle = \langle x|\left[X, \frac{\hat{K}}{1-\hat{K}}\right]|y\rangle = \langle x|\frac{1}{1-\hat{K}}[X,\hat{K}]\frac{1}{1-\hat{K}}|y\rangle \tag{5.34}$$

From the definition of the kernel $\hat{K}$ (5.31), and then (5.34) reads

$$\tilde{K}(x,y) = \frac{q(x)p(y) - q(y)p(x)}{x-y} \tag{5.35}$$

Since $q(x)$ and $p(x)$ depend upon the interval $[a, b]$, we denote them now $q(b, a; x)$ and $p(b, a; x)$ to emphasize their dependence on the interval. Consider now the derivative of $q(b, a; x = b)$, with respect to b ; it consists of two terms

$$\frac{\partial q(b,a;b)}{\partial b} = \langle b|D\frac{1}{1-\hat{K}}|\phi\rangle + \langle b|\frac{1}{1-\hat{K}}\frac{\partial \hat{K}}{\partial b}\frac{1}{1-\hat{K}}|\phi\rangle \tag{5.36}$$

where D is the derivative operator. $\langle x|D|f\rangle = f'(x)$. From the definition of $\hat{K}$,

$$\frac{\partial \hat{K}(x,y)}{\partial b} = K(x,y)\delta(b-y) = \langle x|K|b\rangle\langle b|y\rangle \tag{5.37}$$

Then the second term of (5.36) is $\tilde{K}(b,b)q(b)$. The first term is

$$\begin{aligned}\langle b|D\frac{1}{1-\hat{K}}|\phi\rangle &= p(b) + \langle b|\left[D, \frac{1}{1-\hat{K}}\right]|\phi\rangle \\ &= p(b) + \langle b|\frac{1}{1-\hat{K}}[D,\hat{K}]\frac{1}{1-\hat{K}}|\phi\rangle\end{aligned} \tag{5.38}$$

The commutator is

$$\langle x|[D,\hat{K}]|y\rangle = \left(\frac{\partial}{\partial x} + \frac{\partial}{\partial y}\right)\hat{K}(x,y) = K(x,y)[\delta(y-a) - \delta(y-b)] \tag{5.39}$$

and it reads to

$$[D,\hat{K}] = K|a\rangle\langle a| - K|b\rangle\langle b| \tag{5.40}$$

Thus (5.36) becomes

$$\frac{\partial q(b,a;b)}{\partial b} = p(b,a;b) + \tilde{K}(b,a)q(b,a;a) \tag{5.41}$$

Similarly, for $p(b)$,

$$\frac{\partial p(b, a; b)}{\partial b} = -q(b, a; b) + \tilde{K}(b, a)p(b, a; a) \tag{5.42}$$

Let us now take a symmetric interval $(-b, b)$ with $a = -b$ and denote $q(b, -b; b) = Q(b)$. Then

$$\frac{dQ(b)}{db} = \frac{\partial q(b, a; b)}{\partial b}|_{a=-b} - \frac{\partial q(b, a; b)}{\partial a}|_{a=-b}. \tag{5.43}$$

The last term involves

$$\frac{\partial q(b, a; b)}{\partial a} = \langle b|\frac{1}{1-\hat{K}}\left(\frac{\partial \hat{K}}{\partial a}\right)\frac{1}{1-\hat{K}}|\phi\rangle = -\tilde{K}(b, a)q(b, a; a) \tag{5.44}$$

A nonlinear differential equation follows

$$\dot{Q}(b) = P(b) + 2\tilde{K}(b, -b)Q(-b) = P(b)\left(1 - \frac{2Q^2}{b}\right) \tag{5.45}$$

with $Q(-b) = -Q(b), P(-b) = P(b)$, and $\tilde{K}(b, -b) = \frac{Q(b)P(b)}{b}$. Similarly, for $P(b)$,

$$\dot{P}(b) = Q\left(\frac{2P^2}{b} - 1\right). \tag{5.46}$$

The logarithmic derivative of the level spacing probability $E(s)$ is given by

$$\tilde{K}(b, b) = P^2 + Q^2 - \frac{2P^2Q^2}{b} \tag{5.47}$$

If we replace b by $b = \frac{1}{2}s$, the coupled equations are

$$\frac{dQ}{ds} = \frac{P}{2}\left(1 - \frac{4}{s}Q^2\right), \quad \frac{dP}{ds} = \frac{Q}{2}\left(\frac{4P^2}{s} - 1\right). \tag{5.48}$$

For small s, the solution of this system is

$$Q = \frac{s}{2} - \frac{s^3}{48} + O(s^5), \qquad P = 1 + s + \frac{7}{8}s^2 + O(s^3), \tag{5.49}$$

and

$$H(s) = \tilde{K}(b, b) = 1 + s + s^2 + O(s^3), \tag{5.50}$$

$$E(s) = \exp\left[-\int_0^s H(s')ds'\right] = 1 - s + O(s^4). \tag{5.51}$$

5.3 Universality Classes at an Edge and at a Gap Closure

1. Edge distribution

Near the edge of the spectrum, in scale at which the eigenvalues are within a distance $N^{-1/3}$ of the edge, an Airy kernel governs the spacing distribution. It is defined as

$$K(x,y) = \frac{A_i(x)A_i'(y) - A_i'(x)A_i(y)}{x-y} \tag{5.52}$$

where $A_i(x)$ is an Airy function which replaces the $\phi(x)$ of (5.30). In the Airy case it satisfies $\phi''(x) = x\phi(x)$. The probability $E(s)$ of the previous section is now considered for the interval $[s, \infty]$.

As in the sine kernel case, the Fredholm resolvent $\tilde{K}(a,b)$ is given by

$$\tilde{K}(a,b) = \frac{q(a)p(b) - p(a)q(b)}{a-b} \tag{5.53}$$

From the differential equation $\phi''(x) = x\phi(x)$, one obtains

$$\left(\frac{\partial}{\partial x} + \frac{\partial}{\partial y}\right) K(x,y) = -\phi(x)\phi(y) \tag{5.54}$$

$$[D,K] = -|\phi\rangle\langle\phi|\Theta + K|a\rangle\langle a| - K|b\rangle\langle b| \tag{5.55}$$

where Θ is defined as

$$\Theta(y) = \theta(y-a)\theta(b-y) \tag{5.56}$$

with the Heaviside function $\theta(x)$. As in the sine case,

$$\begin{aligned} \frac{\partial q(b)}{\partial b} &= p(b) - q(b)u + \tilde{K}(b,a)q(a) \\ \frac{\partial p}{\partial b} &= bq(b) + up(b) - 2q(b)v + \tilde{K}(b,a)p(a) \end{aligned} \tag{5.57}$$

where $u = \langle\phi|q\rangle$, $v = \langle\phi|p\rangle$. The Fredholm resolvent $\tilde{K}(b,b)$, which acts as a Hamiltonian $H(b)$, is now

$$\begin{aligned} H(b) &= \tilde{K}(b,b) = p^2(b) - bq^2(b) - 2up(b)q(b) + 2q^2(b)v \\ &+ \frac{1}{b-a}[q(b)p(a) - p(b)q(a)](q(a)p(b) - p(a)q(b)) \end{aligned} \tag{5.58}$$

and indeed it has the Hamiltonian property

$$\frac{\partial H(b)}{\partial p(b)} = 2\frac{\partial q(b)}{\partial b}, \qquad \frac{\partial H(b)}{\partial q(b)} = -2\frac{\partial p(b)}{\partial b}. \tag{5.59}$$

The derivative of this Hamiltonian is

$$\frac{dH(b)}{db} = -q^2(b) + \frac{[q(b)p(a) - p(b)q(a)]^2}{(b-a)^2} \tag{5.60}$$

If we now set $a = -\frac{s}{2}$,

$$\begin{aligned} \frac{\partial q(a)}{\partial a} &= p(a) - q(a)u - \tilde{K}(a,b)q(b), \\ \frac{\partial p(a)}{\partial a} &= aq(a) + up(a) - 2q(a)v - \tilde{K}(a,b)p(b) \end{aligned} \tag{5.61}$$

The Hamiltonian $H(a)$ becomes

$$H(a) = p^2(a) - aq^2(a) - 2up(a)q(a) + 2q^2(a)v + \tilde{K}(a,b)[q(a)p(b) - p(a)q(b)] \tag{5.62}$$

Taking now a large b limit, $\tilde{K}(a,b)$ can be neglected. From the relation

$$u^2 - 2v = q^2 \tag{5.63}$$

we obtain

$$\frac{d^2q(a)}{da^2} = aq(a) + 2q^3(a) \tag{5.64}$$

which is a Painlevé equation of type II.

The solution for large s behaves as $q(a) \sim \pm\frac{1}{2}\sqrt{s}$ and $H(a) \sim \frac{1}{16}s^2$. Therefore the large s spacing distribution behaves as

$$E(s) \sim \exp\left[-\frac{1}{96}s^3\right] \tag{5.65}$$

2. Gap closure

We consider now a matrix source A which possesses only two eigenvalues $a_\gamma = a$ and $a_\gamma = -a$, both of them $N/2$ times degenerate. In the large N limit, the resolvent (Green function) satisfies [16]

$$\begin{aligned} G(z) &= = \frac{1}{N}\sum_\gamma \frac{1}{z - a_\gamma - G(z)} \\ &= \frac{1}{2}\left(\frac{1}{z-a-G(z)} + \frac{1}{z+a-G(z)}\right) \end{aligned} \tag{5.66}$$

This reads

$$G^3 - 2zG^2 + z^2G = z + (a^2-1)G \tag{5.67}$$

Therefore, for $a^2 = 1$, $G \sim z^{\frac{1}{3}}$ for z small, and the density of state $\rho(x)$ behaves as $\rho(x) \sim x^{\frac{1}{3}}$ near the origin. For $|a| > 1$, a gap appears around $x = 0$. Therefore, the case of a source $a_\gamma = \pm 1$ corresponds to a new universality case [21]. The kernel is now described by a Pearcey integral [9, 21, 22, 110, 115, 127], i.e. a higher Airy function, The expression of the kernel $K(\lambda_1, \lambda_2)$ with an external source $a_\gamma = \pm a$, is given in (3.25). The gap closure scaling is governed by the large N- limit

$$\lambda_1 = N^{-\frac{3}{4}}x, \quad \lambda_2 = N^{-\frac{3}{4}}y$$
$$K(x, y) = N^{\frac{1}{4}}K_N(N^{\frac{3}{4}}\lambda_1, N^{\frac{3}{4}}\lambda_2) \tag{5.68}$$

The density of state $\rho(\lambda)$ is given by $K(\lambda, \lambda)$, and the derivative of $\rho(\lambda)$ is

$$\frac{1}{N}\frac{\partial}{\partial\lambda}\rho(\lambda) = -\phi(\lambda)\psi(\lambda) \tag{5.69}$$

with

$$\phi(\lambda) = \int_{-\infty}^{+\infty} \frac{dt}{2\pi} e^{-\frac{N}{2}t^2 + \frac{N}{2}\ln(a^2+t^2) - Nit\lambda} \tag{5.70}$$

$$\psi(\lambda) = \oint \frac{du}{2i\pi} e^{-\frac{N}{2}u^2 - \frac{N}{2}\ln(a^2-u^2) + Nu\lambda} \tag{5.71}$$

In the large N limit, with $a^2 = 1$, it reduces to

$$\phi(\lambda) = \int_{-\infty}^{+\infty} \frac{dt}{2\pi} e^{-\frac{N}{4}t^4 - Nit\lambda} \tag{5.72}$$

$$\psi(\lambda) = \int_c \frac{du}{2i\pi} e^{\frac{N}{4}u^4 + Nu\lambda} \tag{5.73}$$

These two functions satisfy

$$\phi'''(x) = x\phi(x), \quad \psi'''(x) = -x\psi(x). \tag{5.74}$$

This new kernel $K(x, y)$ has a form slightly different from (5.30) [21].

Theorem 5.3 *When the external source $a_\gamma = \pm 1$, the gap closure kernel (3.25) is given by*

$$K(x, y) = \frac{\phi'(x)\psi'(y) - \phi''(x)\psi(y) - \phi(x)\psi''(y)}{x - y}. \tag{5.75}$$

Proof of Theorem 5.3 The kernel is given as

$$K_N(\lambda,\mu) = (-1)^{N-1}\int\frac{dt}{2\pi}\oint\frac{du}{2i\pi}\prod_{\gamma=1}^{N}\left(\frac{a_\gamma - it}{u-a_\gamma}\right)\frac{1}{u-it}$$
$$\times\, e^{-\frac{N}{2}(u^2+t^2)-Nit\lambda+Nu\mu} \tag{5.76}$$

The derivative of the kernel,

$$\frac{\partial}{\partial z}K(x+z,y+z) = -\phi(x+z)\psi(y+z) \tag{5.77}$$

From this equation, one finds

$$(x-y)\frac{\partial}{\partial z}K(x+z,y+z) = -[(x+z)-(y+z)]\phi(x+z)\psi(y+z)$$
$$= -\Big(\phi'''(x+z)\psi(y+z) + \phi(x+z)\psi'''(y+z)\Big)$$
$$= -\frac{\partial}{\partial z}\Big(\phi''(x+z)\psi(y+z) + \phi(x+z)\psi''(y+z) - \phi'(x+z)\psi'(y+z)\Big) \tag{5.78}$$

The integration over z gives

$$(x-y)K(x+z,y+z)$$
$$= -\Big(\phi''(x+z)\psi(y+z) + \phi(x+z)\psi''(y+z) - \phi'(x+z)\psi'(y+z)\Big)$$
$$+C(x,y). \tag{5.79}$$

Since the kernel satisfies

$$\left(\frac{\partial}{\partial x}+\frac{\partial}{\partial y}\right)K(x,y) = -\phi(x)\psi(y) \tag{5.80}$$

using (5.74) and setting $z=0$ one finds

$$\left(\frac{\partial}{\partial x}+\frac{\partial}{\partial y}\right)C(x,y) = 0 \tag{5.81}$$

This leads to

$$C(x,y) = C(x-y) \tag{5.82}$$

When $y=0$ and $z=0$, from (5.79) one obtains

$$xK(x,0) = -(\phi''(x)\psi(x) + \phi(x)\psi''(x) - \phi'(x)\psi'(x)) + C(x) \tag{5.83}$$

For $x \to 0$, one finds $\lim_{x\to 0} C(x) = 0$, since $\phi'(0) = \psi(0) = \psi''(0) = 0$. Therefore, the kernel is expressed as stated in Theorem 5.3.2. □

When the external source a is close to ± 1, the scaling region of this new universality class, described by a parameter α [22] defined by

$$a^2 = 1 + \frac{2}{\sqrt{N}}\alpha. \tag{5.84}$$

Then the previous two functions ϕ and ψ are modified as

$$\phi(\lambda) = \int_{-\infty}^{\infty} \frac{dt}{2\pi} e^{-\frac{1}{4}t^4 - \alpha t^2 + it\lambda} \tag{5.85}$$

which satisfies

$$\phi''' - 2\alpha\phi' - \lambda\phi = 0 \tag{5.86}$$

The $\psi(\lambda)$ is also modified as

$$\psi''' - 2\alpha\psi' + \lambda\psi = 0 \tag{5.87}$$

Following the previous analysis for $a = \pm 1$, the kernel is then expressed as

Proposition 5.3.1 *The kernel $K(x, y)$ in the scaling region for $a^2 = 1 + \frac{2}{\sqrt{N}}\alpha$ becomes*

$$K(x, y) = \frac{\phi'(x)\psi'(y) - \phi''(x)\psi(y) - \phi(x)\psi''(y) + 2\alpha\phi(x)\psi(y)}{x - y} \tag{5.88}$$

Proof of Proposition 5.3.1 Following the previous analysis for Theorem 5.3, we have obtained the kernel governing the scaling region $a^2 = 1 + \frac{2}{\sqrt{N}}\alpha$ [22]. □
The large λ behavior of $\phi(x)$, for fixed α, is obtained by a saddle point method. Changing t to $\lambda^{1/3}t$, one finds that the term αt^2 becomes negligible compared with the other terms of order $\lambda^{4/3}$. Then one obtains the large x behavior of $\rho(x)$ as $x^{1/3}$ as before.

The level spacing probability $p(s)$ for this gap closure case is also studied by the Fredholm theory as in the Airy kernel case. The level spacing function $E(s)$, the probability that there is no eigenvalue inside the interval $\left(-\frac{s}{2}, \frac{s}{2}\right)$ centered around the singular point $s = 0$, is given by the Fredholm determinant

$$E(a, b) = \det[1 - \hat{K}] = \sum_{n=0}^{\infty} \frac{(-1)^n}{n!} \int_a^b \cdots \int_a^b \prod_{l=1}^{n} dx_l \times \det[K_{i,x_j}]_{i,j=1,\ldots,n} \tag{5.89}$$

where the interval (a, b) is $\left(-\frac{s}{2}, \frac{s}{2}\right)$. The kernel $K(x, y)$ in (5.75) is not symmetric, since two functions $\phi(x)$ and $\psi(x)$ are different. Therefore, the kernel $\hat{K}(x, y)$ is defined by the restriction of K to the interval:

$$\hat{K}(x, y) = K(x, y)\theta(y - a)\theta(b - y) = K(x, y)\Theta(y), \tag{5.90}$$

where $\theta(x)$ is the Heaviside function, and for convenience the notation Θ stands for

$$\Theta(y) = \theta(y - a)\theta(b - y). \tag{5.91}$$

From

$$\log E(a, b) = \mathrm{tr}\log(1 - \hat{K}), \tag{5.92}$$

one obtain

$$\frac{\partial \log E(a, b)}{\partial b} = -\mathrm{tr}\left(\frac{1}{1 - \hat{K}} \frac{\partial \hat{K}}{\partial b}\right). \tag{5.93}$$

From (5.90), one finds

$$\frac{\partial \hat{K}(x, y)}{\partial b} = K(x, b)\delta(y - b). \tag{5.94}$$

The Fredholm resolvent $\tilde{K}(b, b)$ is defined as

$$\tilde{K} = \frac{\hat{K}}{1 - \hat{K}}. \tag{5.95}$$

From (5.93) and (5.94), one obtains

$$\frac{\partial \log E(a, b)}{\partial b} = -\tilde{K}(b, b), \tag{5.96}$$

and similarly

$$\frac{\partial \log E(a, b)}{\partial a} = \tilde{K}(a, a), \tag{5.97}$$

When the interval (a, b) is symmetric, namely it is $\left(-\frac{s}{2}, \frac{s}{2}\right)$, one obtains

$$\frac{d\log E(s)}{ds} = \frac{1}{2}\left(\frac{\partial}{\partial b} - \frac{\partial}{\partial a}\right)\log E(s)|_{b=-a=\frac{s}{2}} = -\tilde{K}\left(\frac{s}{2}, \frac{s}{2}\right). \tag{5.98}$$

This leads to the following proposition,

Proposition 5.3.2

$$E(s) = \exp\left[-\int_0^s \tilde{K}\left(\frac{s'}{2}, \frac{s'}{2}\right) ds'\right]. \tag{5.99}$$

To solve $\tilde{K}$, six functions are introduced. These are obtained by the action of $(1-\hat{K})^{-1}$ on the two functions ϕ and ψ, and on the their first two derivatives.

$$q_0(x) = (1-\hat{K})^{-1}\phi(x) = \phi(x) + \int_a^b K(x,y)\phi(y)dy + \int_a^b \int_a^b K(x,y)K(y,z)\phi(z)dydz + \cdots . \tag{5.100}$$

It is convenient to use Dirac's notations

$$\begin{aligned} q_0(b,a;x) &= \langle x|\frac{1}{1-\hat{K}}|\phi\rangle \\ q_n(b,a;x) &= \langle x|\frac{1}{1-\hat{K}}|\phi^{(n)}\rangle, \quad (n=1,2) \\ p_n(b,a;x) &= (-1)^{n-1}\langle\psi^{(2-n)}|\frac{1}{1-\hat{L}}|x\rangle, \quad (n=0,1,2). \end{aligned} \tag{5.101}$$

where

$$\hat{L}(y,x) = \Theta(y)K(y,x). \tag{5.102}$$

When $x = b$ and $a = -b$, these six functions q_n and p_n become functions of the single variable b. They are denoted as

$$Q_n(b) = q_n(b,-b;b), \quad P_n(b) = p_n(b,-b;b). \tag{5.103}$$

These six functions satisfy the following differential equation of b, (a dot means taking the derivative with respect to b).

$$\begin{aligned} \dot{Q}_0 &= Q_1 + \frac{2}{b}Q_1P_1Q_0, \\ \dot{Q}_1 &= Q_2 - \frac{2}{b}Q_1^2P_1 - Q_0u, \\ \dot{Q}_2 &= bQ_0 + \frac{2}{b}Q_1P_1Q_2 - Q_1v, \\ \dot{P}_0 &= -bP_2 - \frac{2}{b}Q_1P_1P_0 + P_1u, \\ \dot{P}_1 &= -P_0 + \frac{2}{b}Q_1P_1^2 + P_2v, \\ \dot{P}_2 &= -Pq_n(b,a)_1 - \frac{2}{b}Q_1P_1P_2. \end{aligned} \tag{5.104}$$

where the two auxiliary functions u and v are defined as

$$u = \langle\psi|q_1\rangle, \quad v = \langle\psi'|q_0\rangle \tag{5.105}$$

Using operator notation,

$$[X, K] = |\phi'\rangle\langle\psi'| - |\phi''\rangle\langle\psi| - |\phi\rangle\langle\psi''| \tag{5.106}$$

and

$$\begin{aligned}(x-y)\tilde{K} &= \langle x|\frac{1}{1-\hat{K}}[X, \hat{K}]\frac{1}{1-\hat{K}}|y\rangle \\ &= q_1(x)p_1(x) - q_2(x)p_0(y) - q_0(x)p_2(y)\end{aligned} \tag{5.107}$$

The derivatives of $q_n(b, a; b)$ for fixed a is

$$\begin{aligned}\frac{\partial q_n(b, a; b)}{\partial b} &= \langle b|D\frac{1}{1-\hat{K}}|\phi^{(n)}\rangle + \langle b|\frac{1}{1-\hat{K}}\left(\frac{\partial \hat{K}}{\partial b}\right)\frac{1}{1-\hat{K}}|\phi^{(n)}\rangle \\ &= q_{n+1}(b, a; b) + \langle b|\frac{1}{1-\hat{K}}[D, \hat{K}]\frac{1}{1-\hat{K}}|\phi^{(n)}\rangle \\ &+ \langle b|\frac{\hat{K}}{1-\hat{K}}|b\rangle\langle b|\frac{1}{1-\hat{K}}|\phi^{(n)}\rangle\end{aligned} \tag{5.108}$$

where D is the derivative operator: $\langle x|D|f\rangle = f'(x)$.
Then $[D, \hat{K}]$ reads

$$\langle x|[D, \hat{K}]|y\rangle = \left(\frac{\partial}{\partial x} + \frac{\partial}{\partial y}\right)K(x, y) + \langle x|K|a\rangle\langle a|y\rangle - \langle x|K|b\rangle\langle b|y\rangle. \tag{5.109}$$

The first term is $-\phi(x)\psi(y)$. Then, since

$$[D, \hat{K}] = -|\phi\rangle\langle\psi|\Theta + K|a\rangle\langle a| - K|b\rangle\langle b| \tag{5.110}$$

one finds

$$\frac{\partial q_n(b, a; b)}{\partial b} = q_{n+1} + \tilde{K}(b, a)q_n(a) - q_0(b)\langle\psi|q_n\rangle \tag{5.111}$$

and

$$\begin{aligned}q_3 &= \langle x|\frac{1}{1-\hat{K}}X|\phi\rangle \\ &= \langle x|X\frac{1}{1-\hat{K}}|\phi\rangle + \langle x|\left[\frac{1}{1-\hat{K}}, X\right]|\phi\rangle \\ &= xq_0(x) + \langle x|\frac{1}{1-\hat{K}}[\hat{K}, X]\frac{1}{1-\hat{K}}|\phi\rangle \\ &= xq_0 - v_2q_1 + u_1q_2 + v_3q_0\end{aligned} \tag{5.112}$$

where $u_1 = \langle\psi|q_0\rangle$, $v_2 = \langle\psi'|q_0\rangle$ and $v_3 = \langle\psi''|q_0\rangle$.

For $p_n(x)$, one obtains similarly the following equations.

$$p_n(x) = (-1)^{n-1}\langle\psi^{(2-n)}|\frac{1}{1-\hat{L}}|x\rangle, \tag{5.113}$$

with $\hat{L}(y,x) = \Theta(y)K(y,x)$. The derivative operator D acts as

$$[D,\hat{L}] = -\Theta|\phi\rangle\langle\psi| + |a\rangle\langle a|K - |b\rangle\langle b|K \tag{5.114}$$

in which Θ is a local operator defined by

$$\langle y|\Theta|y'\rangle = \delta(y-y')\theta(y-a)\theta(b-y) \tag{5.115}$$

Therefore, one obtains

$$\frac{\partial p_n(b)}{\partial b} = -p_{n-1}(b) - p_0(b)\langle\psi^{(2-n)}|q_0\rangle + p_n(a)\tilde{K}(a,b) \tag{5.116}$$

with p_{-1}, which is

$$\begin{aligned} p_{-1}(x) &= -\langle\psi'''|\frac{1}{1-\hat{L}}|x\rangle \\ &= xp_2(x) - \langle\psi|\frac{1}{1-\hat{L}}[x,\hat{L}]\frac{1}{1-\hat{L}}|x\rangle \\ &= -xp_0(x) - p_1(x)\langle\psi|q_1\rangle - p_2(x)\langle\psi|q_2\rangle - p_0(x)\langle\psi|q_0\rangle, \end{aligned} \tag{5.117}$$

where

$$\begin{aligned} \langle y|[X,\hat{L}]|x\rangle &= (y-x)\Theta(y)K(y,x) \\ &= \Theta(y)\Big(|\phi'\rangle\langle\psi'| - |\phi''\rangle\langle\psi| - |\phi\rangle\langle\psi''|\Big). \end{aligned} \tag{5.118}$$

Noting that $\phi(x)$ is an even function of x and $\psi(x)$ is an odd function, one finds $u_1 = v_3 = 0$. Also $\langle\psi|q_2\rangle = \langle\psi|q_0\rangle = \langle\psi''|q_0\rangle = 0$. Non vanishing quantities are $u_2 = \langle\psi|q_1\rangle$ and $v_2 = \langle\psi'|q_0\rangle$, which are denoted simply as u and v.

The derivative of this u with respect to b becomes

$$\dot{u} = -2P_2(b)Q_1(b) \tag{5.119}$$

Since

$$\frac{\partial\hat{K}}{\partial b} = K(x,y)\delta(y-b) = K|b\rangle\langle b|, \tag{5.120}$$

one obtains

$$\frac{\partial u}{\partial b} = \langle \psi | \Theta \frac{1}{1-\hat{K}} \frac{\partial \hat{K}}{\partial b} \frac{1}{1-\hat{K}} |\phi'\rangle \psi(b) q_1(b). \tag{5.121}$$

This leads to

$$\frac{\partial u}{\partial b} = -p_2(b) q_1(b). \tag{5.122}$$

The derivative with respect to a is similar.

$$\frac{\partial u}{\partial a} = p_2(a) q_1(a). \tag{5.123}$$

Then, by putting $a = -b$, one obtains

$$\dot{u} = \frac{\partial u}{\partial b}|_{a=-b} - \frac{\partial u}{\partial a}|_{a=-b} = -2P_2(b) Q_1(b). \tag{5.124}$$

which is (5.119). For v, one obtains similarly,

$$\dot{v} = 2P_1 Q_0. \tag{5.125}$$

From the derivatives of u and v, one finds

$$u + v = -2P_2 Q_0 \tag{5.126}$$

and

$$\dot{Q}_0 = Q_1 \left(1 + \frac{\dot{v}}{b}\right), \quad \dot{P}_2 = -P_1 \left(1 - \frac{\dot{u}}{b}\right) \tag{5.127}$$

From (5.126),(5.119),(5.125) and (5.104) one verifies the closed coupled equations for u and v.

$$\begin{aligned} -2P_2 Q_2 &= \ddot{u} - \frac{\dot{u}\dot{v}}{u+v} + \frac{2}{b} \frac{\dot{v}\dot{u}^2}{u+v} + u(u+v), \\ -2Q_0 P_0 &= \ddot{v} - \frac{\dot{u}\dot{v}}{u+v} - \frac{2}{b} \frac{\dot{u}\dot{v}^2}{u+v} + v(u+v) \end{aligned} \tag{5.128}$$

Taking the derivatives of these two equations, one obtains coupled equations for u and v

$$\begin{aligned} \frac{d^3 u}{db^3} &+ \left(\frac{2\dot{u}}{b} - 1\right) \Big[b(u+v) + \frac{1}{u+v} (\ddot{v}\dot{u} + 2\dot{v}\ddot{u}) \\ &- \frac{\dot{u}\dot{v}}{(u+v)^2} (2\dot{v} + \dot{u}) \Big] - \frac{2\dot{v}(\dot{u})^2}{b^2(u+v)} = 0 \end{aligned}$$

$$\frac{d^3v}{db^3} + \left(\frac{2\dot{v}}{b} + 1\right)\left[b(u+v) - \frac{1}{u+v}(\ddot{u}\dot{v} + 2\dot{u}\ddot{v}) + \frac{\dot{u}\dot{v}}{(u+v)^2}(2\dot{u} + \dot{v})\right] + \frac{2\dot{u}(\dot{v})^2}{b^2(u+v)} = 0. \tag{5.129}$$

In the large b limit, the solutions of these equations are

$$u = \frac{b^2}{4} - \frac{1}{2}\left(\frac{1}{4}\right)^{\frac{1}{3}} b^{\frac{2}{3}} + \cdots,$$
$$v = -\frac{b^2}{4} - \frac{1}{2}\left(\frac{1}{4}\right)^{\frac{1}{3}} b^{\frac{2}{3}} + \cdots, \tag{5.130}$$

The kernel $\tilde{K}(b, b)$ is expressed as

$$\tilde{K}(b,b) = bP_2Q_0 + Q_2P_1 + Q_1P_0 - uP_1Q_0 - vP_2Q_1 - \frac{1}{2b}(P_1^2Q_1^2 - Q_2P_2 - Q_0P_0)^2. \tag{5.131}$$

This kernel $\tilde{K}(b, b)$ becomes again a Hamiltonian

$$H(b) = \tilde{K}(b,b)$$
$$\dot{Q}_n = \frac{\partial H}{\partial P_n}, \quad \dot{P}_n = -\frac{\partial H}{\partial Q_n} \tag{5.132}$$

which indeed provide the differential equations (5.104). Thus, one finds

$$\frac{dH(b)}{db} = Q_0P_2 + \frac{2}{b^2}P_1^2Q_1^2 = -\frac{u+v}{2} + \frac{(\dot{u}\dot{v})^2}{2b^2(u+v)^2}. \tag{5.133}$$

For large b, from (5.130) one obtains

$$\frac{dH(b)}{db} = 5 \times 2^{-\frac{11}{3}} b^{\frac{2}{3}} \tag{5.134}$$

which leads $H(b) \sim 3 \times 2^{-\frac{11}{3}} b^{\frac{5}{3}} = 3 \times 2^{-\frac{16}{3}} s^{\frac{5}{3}}$. Thus $E(s)$ behaves in the large s limit as

$$E(s) = D\exp\left[-\int_0^s H(s')ds'\right] \sim \exp[-9 \times 2^{-\frac{25}{3}} s^{\frac{8}{3}}], \tag{5.135}$$

where D is a constant. The exponent $\frac{8}{3}$ agrees with the exponent β of the density of state $\rho(x)$, namely $\frac{8}{3} = 2\beta + 2$, in which $\beta = \frac{1}{3}$ is the exponent governing the density of states near the gap closure point.

In this closure gap singularity, the n-point function $U(\sigma_1, \ldots, \sigma_n)$ becomes the generating function of the intersection numbers of p-spin curves (p=3), which will be discussed in Sect. 7.2.

3. Higher edge singularities

Higher edge singularities may be obtained by an appropriate tuning of the external source eigenvalues a_γ. The next singularity is obtained with a source matrix possessing three distinct eigenvalues, with $a_\gamma = a_1, a_2, a_3$, each one $N/3$ times degenerate. The new singularity corresponds to a choice of three numbers which satisfy

$$\frac{1}{a_1^2} + \frac{1}{a_2^2} + \frac{1}{a_3^2} = 3,$$
$$\frac{1}{a_1^n} + \frac{1}{a_2^n} + \frac{1}{a_3^n} = 0, \ (n = 3, 4) \tag{5.136}$$

The a_α $(\alpha = 1, 2, 3)$ are the solution of the cubic equation

$$1 + \beta x + \gamma x^2 + \delta x^3 = 0 \tag{5.137}$$

with

$$\beta^2 = 9 \pm 3\sqrt{6}, \gamma = \frac{1}{2}(\beta^2 - 3), \delta = \frac{3}{2\beta}(\beta^2 - 3) \tag{5.138}$$

The solutions for the a_i are [29]

$$(a_1, a_2, a_3) = (\pm 0.52523, \pm 0.41127 \pm 0.46403i, \pm 0.41127 \mp 0.46403i),$$
$$(a_1, a_2, a_3) = (\pm 1.0076, \mp 0.71801 \pm 0.33908i, \mp 0.71801 \mp 0.33908i) \tag{5.139}$$

Both solutions give a density of state, which behaves in the vicinity of the edge as $\rho(\lambda) \sim \lambda^{\frac{1}{4}}$ Indeed in the large N limit, the resolvent satisfies

$$G(z) = \frac{1}{3}\left(\frac{1}{z - a_1 - G(z)} + \frac{1}{z - a_2 - G(z)} + \frac{1}{z - a_3 - G(z)}\right) \tag{5.140}$$

This corresponds to a singularity governed by a higher Airy function, which satisfies

$$\phi''''(x) = x\phi(x) \tag{5.141}$$

The singularity corresponds to $p = 4$ of p-spin curves, and will be discussed later in Sect. 7.2.

The next singularity $p = 5$ is obtained with four distinct a_i , $\frac{N}{4}$ times degenerate and

$$G(z) = \frac{1}{4}\left(\frac{1}{z - a_1 - G(z)} + \frac{1}{z - a_2 - G(z)} + \frac{1}{z - a_3 - G(z)} + \frac{1}{z - a_4 - G(z)}\right) \tag{5.142}$$

provided the *ai* satisfy

$$\begin{aligned} &\frac{1}{a_1^2} + \frac{1}{a_2^2} + \frac{1}{a_3^2} + \frac{1}{a_4^2} = 4, \\ &\frac{1}{a_1^n} + \frac{1}{a_2^n} + \frac{1}{a_3^n} + \frac{1}{a_4^n} = 0, \quad (n = 3, 4, 5). \end{aligned} \tag{5.143}$$

The solutions (up to 4 digits) belong to three different classes [29]

$$\begin{aligned} (a_1, a_2, a_3, a_4) &= (\rho + i\xi, \rho - i\xi, -\rho + i\xi, -\rho - i\xi) \\ (\rho &= \pm 0.7769, \xi = \pm 0.3218) \\ (a_1, a_2, a_3, a_4) &= \pm(0.6249, -1.014, 0.5336 + i0.4735, 0.5336 - i0.4735) \\ (a_1, a_2, a_3, a_4) &= \pm(0.2806 + i0.5117, 0.2806 - i0.5117, 0.4337 + i0.1589, \\ &\qquad 0.4337 - i0.1589) \end{aligned} \tag{5.144}$$

All these cases give a closing gap singularity for the density of state $\rho(\lambda) \sim \lambda^{\frac{1}{5}}$. This gap closure singularity is related to the $p = 5$ spin curve of Chap. 7. The kernel $K(x, y)$ has been studied in [21]. It is given by

$$K(\lambda, \mu) = -\int_{-\infty}^{\infty} \frac{dt}{2\pi} \oint \frac{du}{2i\pi} \frac{1}{u - it} e^{-\frac{1}{3}(t^6 + u^6) - it\lambda + u\mu}. \tag{5.145}$$

The scaling limit is obtained by scaling t and u as $N^{-\frac{1}{6}}$, and λ and μ as $N^{-\frac{5}{6}}$. The scaled kernel $K(x, y)$, which is a generalization of a gap closure kernel (5.75), is

$$\begin{aligned} K(x, y) = \frac{2}{x - y}[&\phi''''(x)\psi(y) - \phi'''(x)\psi'(y) + \phi''(x)\psi''(y) \\ &- \phi'(x)\psi'''(y) + \phi(x)\psi''''(y)], \end{aligned} \tag{5.146}$$

with

$$\begin{aligned} \phi(x) &= \int_{-\infty}^{\infty} \frac{dt}{2\pi} e^{-\frac{1}{3}t^6 + itx} \\ \psi(x) &= \int_c \frac{du}{2i\pi} e^{-\frac{1}{3}u^6 + ux} \end{aligned} \tag{5.147}$$

These two functions satisfy

$$\frac{d^5}{dx^5}\phi(x) = -\frac{1}{2}x\phi(x),$$
$$\frac{d^5}{dx^5}\psi(x) = \frac{1}{2}x\psi(x) \tag{5.148}$$

and one can repeat the same analysis leading to a Hamiltonian system which governs the level spacing distribution.

5.4 Distribution of Zeros of Riemann's Zeta Function

It is well-known that the distribution of the zeros of Riemann's zeta function on the critical line are in total agreement with the spacings of eigenvalues of the GUE ensemble of random matrices [100, 105]. It is interesting to compare the moments of this distribution to the average of characteristic polynomials.

Let us begin with the characteristic polynomials and extend the probability measure to non-Gaussian distributions governed by polynomials $V(x)$

$$P(X) = e^{-\frac{N}{2}\mathrm{tr}V(X)} \tag{5.149}$$

For this probability measure, the average of characteristic polynomials of proposition 4.1 for $\lambda_1 = \cdots = \lambda_{2k} = \lambda$ satisfy the following relation in the large N-limit.

Theorem 5.4

$$e^{-NkV(\lambda)}F_{2K}(\lambda,\ldots,\lambda) = (2\pi N\rho(\lambda))^{K^2}e^{-NK}\prod_{l=0}^{K-1}\frac{l!}{(K+l)!} \tag{5.150}$$

where the factor $\prod_{l=0}^{K-1}\frac{l!}{(K+l)!}$ is a universal number, i.e., it is independent of the potential V.

Proof of Theorem 5.4 [24] The proof starts from the expression of $\langle\prod_{l=1}^{k}\det(\lambda_l - X)\rangle$ for distinct λ_l, in the Dyson limit $N(\lambda_i - \lambda_j)$ fixed, which is universal, i.e. independent of the polynomial $V(X)$. One then sets all the $\lambda_l = \lambda$, i.e. zero Dyson limit.

Let us consider first the Gaussian case

$$P(X) = \frac{1}{Z}\exp\left(-\frac{N}{2}\mathrm{tr}X^2\right) \tag{5.151}$$

for Hermitian matrices of size $M \times M$ (it is interesting to disentangle M and N). We return to the expression of the correlation functions $F_K(\lambda_1,\ldots,\lambda_K)$of characteristic polynomials in terms of orthogonal polynomials (4.24). For a Gaussian, the polynomials $p_n(x)$ are Hermite polynomials,

$$H_n(x) = \frac{(-1)^n}{N^n} e^{\frac{1}{2}Nx^2} (\frac{d}{dx})^n e^{-\frac{N}{2}x^2}$$
$$= \frac{(-1)^n n!}{N^n} \oint \frac{dz}{2i\pi} \frac{e^{-N(\frac{1}{2}z^2+xz)}}{z^{n+1}} \tag{5.152}$$

From (4.24) and above integral representation of Hermite polynomial,

$$F_{2K}(\lambda_1, \ldots, \lambda_{2K})$$
$$= \frac{(-1)^K}{\Delta(\lambda_1, \ldots, \lambda_{2K})} \frac{1}{N^{K(2M+2K-1)}} \prod_{l=0}^{2K-1} (M+l)! \oint \prod_{l=1}^{2K} \frac{dz_l}{2i\pi z_l^{M+l}} \exp\left[-\frac{N}{2}\sum_{l=1}^{2K} z_l^2\right]$$
$$\times \det(e^{-N\lambda_a z_b}) \tag{5.153}$$

This formula is exact for finite N and M. We can expand the determinant in (5.153), and keep only one of $(2K)!$ terms, antisymmetrizing instead of integration of z_l. We have

$$F_{2K}(\lambda_1, \ldots, \lambda_{2K}) = \frac{(-1)^K}{\Delta(\lambda_1, \ldots, \lambda_{2K})} \frac{1}{N^{K(2M+2K-1)}} \prod_{l=0}^{2K-1} (M+l)!$$
$$\times \oint \prod_{l=1}^{2K} \frac{dz_l}{2i\pi z_l^{M+2K}} \exp\left[-N \sum_{l=1}^{2K} \left(\frac{z_l^2}{2} + \lambda_l z_l\right)\right] \Delta(z_1, \ldots, Z_{2K}). \tag{5.154}$$

In the large N limit (and $M/N = O(1)$), the integrand has two saddle points for every z_l, namely the two roots $z_\pm$, of $z^2 + \lambda_l z + 1 = 0$, for every z_l independently (i.e. there are 2^{2K} saddle-points). Let us denote λ and x_a, the parameters

$$\lambda = \frac{1}{2K} \sum_{l=1}^{2K} \lambda_l \tag{5.155}$$

and

$$x_a = 2\pi N \rho(\lambda)(\lambda_a - \lambda), \tag{5.156}$$

the scaling variables, with $\sum_{a=1}^{2K} x_a = 0$, The dominant saddle-points in the large N-limit correspond to K roots equal to z_+ and K roots to z_- , i.e. up to a permutation

$$z_l(\lambda_l) = z_+(\lambda_l), \quad l = 1, \ldots, K,$$
$$z_l(\lambda_l) = z_-(\lambda_l), \quad l = K+1, \ldots, 2K \tag{5.157}$$

Noting that in the scaling limit in which all the $z_+(\lambda_l)$ approach $z_+(\lambda)$ and the $z_-(\lambda_l)$ approach $z_-(\lambda)$

$$\frac{\Delta(z_1,\ldots,z_{2K})}{\Delta(\lambda_1,\ldots,\lambda_{2K})} = \left(\frac{dz_+}{d\lambda}\frac{dz_-}{d\lambda}\right)^{\frac{1}{2}K(K-1)} (2i\cos\phi)^{K^2} \prod_{1\le l\le K, K+1\le m\le 2K} \frac{1}{\lambda_l-\lambda_m}$$
$$= (Ni)^{K^2}(2\pi\rho(\lambda))^{K+K^2} \prod_{1\le l\le K, K+1\le m\le 2K} \frac{1}{x_l-x_m} \quad (5.158)$$

in which we have used $\lambda = 2\sin\phi$, so that Wigner's semi-circle corresponds to the density $\rho(\lambda) = \frac{1}{\pi}\cos\phi$. We now compute the contour integral

$$\exp\left[-\frac{N}{2}\sum_{l=1}^{2K} V(\lambda_l)\right] F_{2K}(\lambda_1,\ldots,\lambda_{2K}) = (2\pi N\rho(\lambda))^{K^2}\frac{1}{K!}e^{-NK}$$
$$\times \oint \prod_{\alpha=1}^{K}\frac{du_\alpha}{2\pi}\exp\left[-i\sum_\alpha u_\alpha\right]\prod_{\alpha=1}^{K}\prod_{l=1}^{2K}\frac{1}{u_\alpha-x_l}\Delta(u_1,\ldots,u_K) \quad (5.159)$$

by summing over the 2^{2K} saddle-points and one finds in the limit $\lambda_i = \lambda$, in which $x\to 0$,

$$\exp\left[-NKV(\lambda)\right] F_{2K}(\lambda,\ldots,\lambda)$$
$$= (2\pi N\rho(\lambda))^{K^2}\frac{1}{K!}e^{-NK}\oint \prod_{\alpha=1}^{K}\frac{du_\alpha}{2\pi}\exp\left[-i\sum_\alpha u_\alpha\right]\prod_{\alpha=1}^{K}\frac{1}{u_\alpha^{2K}}\Delta(u_1,\ldots,u_K) \quad (5.160)$$

The last factor in the contour integration reads

$$\oint \prod_{\alpha=1}^{K}\frac{du_\alpha}{2\pi}\exp\left[-i\sum_{\alpha-1}^{K} u_\alpha\right]\prod_{\alpha=1}^{K}\frac{1}{u_\alpha^{2K}}\Delta(u_1,\ldots,u_K) = K!\prod_{l=0}^{K-1}\frac{l!}{(K+l)!} \quad (5.161)$$

The non Gaussian case results in simply a modification of $\rho(\lambda)$. We have thus obtained the Theorem 5.4.

□

Let us now compare with the moments of the zeta function $\zeta(s)$ computed in [40],

$$\frac{1}{T}\int_0^T |\zeta(\frac{1}{2}+it)|^{2k}dt = \frac{1}{T}\int_0^T dt|\sum_{n=1}^{\infty}\frac{d_K(n)}{n^{\frac{1}{2}+it}}|^2 \simeq \gamma_k a_k(\log T)^{k^2} \quad (5.162)$$

where $d_K(n)$ is the Dirichlet coefficient, defined by $d_K(n) = \sum_{n_1\cdots n_K=n} 1$, and γ_k and a_k are

$$\gamma_k = \prod_{l=0}^{k-1} \frac{l!}{(k+l)!}. \tag{5.163}$$

$$a_k = \prod_p \left[(1-p^{-1})^{k^2} \sum_{j=0}^{\infty} \frac{d_k^2(p^j)}{p^j} \right] \tag{5.164}$$

Note that the factor $\log T$ is indeed the asymptotic density of primes as in (5.150) and the factor γ_k is also common to both. The universal coefficient comes from the combinatorics. Similarly we have considered in [24] the integral of a product of ζ functions on the critical line $\prod \zeta(1/2 + it + i\lambda_n)$, integrated from 1 to T over t. Again this is analogous to the contour integration of characteristic polynomials and the combinatoric coefficient (5.150) is also present.

The Riemann zeta function is one of the generalized L-functions, which show the same universal behavior. The ubiquitous occurrence of the coefficient γ_k in (5.150) and in (5.162) is attracting interest.

Chapter 6
Intersection Numbers of Curves

6.1 Kontsevich Airy Matrix Model

Witten [134] conjectured that a generating function of the intersection numbers of the moduli space of curves on a Riemann surface with marked points, is a solution of the KdV hierarchy. Kontsevich [89] has proved this conjecture with the use of an Airy matrix model. In addition it has been realized that matrix models of this type are examples of an exact closed/open strings duality [63].

The definition of intersection numbers on the moduli space of algebraic curves, is mathematically quite involved and we refer the reader for proper definition to ([89, 134]). It deals with the moduli space $\mathcal{M}_g$ of algebraic curves (or Riemann surfaces) on C, of genus g with s marked points. The dimension n of the moduli space for curves of genus g and s marked points is given by $3g-3+s = n$ due to Riemann [118]. To each marked point is associated an integer n_i so that $3g-3+s = n_1 + \cdots + n_s$ for s marked points. The meaning of n_i within a matrix model will be specified below.

The intersection theory of Kontsevich relies upon an expansion of a matrix integral, which produces so called ribbon graphs. The ribbon graphs generated by the double lines Feynman graph expansion of a matrix model make a triangulation of the moduli space. The computation of intersection numbers is reduced to counting the automorphism of Feynman graphs. Algebraic geometrical definition of the intersection numbers are given in the next chapter. We will here show the several intersection numbers, based upon Airy matrix model, through Virasoro equations.

Theorem 6.1 (Kontsevich)

The Airy matrix model is a generating function for the intersection numbers of the moduli space of curves:

$$F(t_0, t_1, \ldots,) = \log\langle \exp\left(\frac{i}{6}\mathrm{tr}M^3\right)\rangle = \langle \exp\left(\sum t_i \tau_i\right)\rangle \tag{6.1}$$

where the matrix average of a function $\langle f(M)\rangle$ is defined with a weight

E. Brézin and S. Hikami, *Random Matrix Theory with an External Source*,
SpringerBriefs in Mathematical Physics, DOI 10.1007/978-981-10-3316-2_6

$$\langle f(M)\rangle = \frac{1}{Z}\int dM f(M)\exp\left[-\frac{1}{2}\mathrm{tr}\Lambda M^2\right] \tag{6.2}$$

where Z is a normalization constant. The expansion parameters t_k are defined as

$$t_k = -\frac{1}{2k+1}\mathrm{tr}\frac{1}{\Lambda^{2k+1}} \tag{6.3}$$

and the generating function $F(t_0, t_1, \ldots,)$ expanded with coefficient τ_i

$$F(t_0, t_1, \ldots,) = \langle \exp\left[\sum_{i=0}^{\infty} t_i\tau_i\right]\rangle = \sum_{k_i}\langle \tau_0^{k_0}\tau_1^{k_1}\cdots\rangle \prod_{i=0}^{\infty}\frac{t_i^{k_i}}{k_i!} \tag{6.4}$$

provide the intersection numbers as coefficients. Note that the size N of the matrix M is absorbed in the trace over powers of the Λ. Therefore, Kontsevich's formula for the intersection numbers is valid for finite N. It is necessary for N to be large enough to avoid algebraic relations between traces of increasing power.

6.2 Evaluation of Intersection Numbers of Curves

The intersection numbers may be computed by several methods. The following three methods have been used successfully:

(i) edge singularity from Gaussian means (2.24), [28, 106]
(ii) replica method based on the replica formula (2.61), [30]
(iii) duality method for the n-point function $U(\sigma_1, \ldots, \sigma_n)$ in presence of a simple external source multiple of the identity $A = 1$, [29, 32].

(i) In (2.24) we have given the Gaussian mean $\frac{1}{N}\langle \mathrm{tr}M^{2k}\rangle$, in the limit $k \to \infty$ and $N \to \infty$, in a scale in which $\frac{k^3}{N^2}$ is finite. The genus one coefficient of $\frac{k^3}{N^2}$ is simply $\frac{1}{12}$, which the $g = 1$ intersection number

$$\langle \tau_{3g-2}\rangle_g = \frac{1}{(12)^g 2^g g!}. \tag{6.5}$$

The scaling limit corresponds to the Tracy–Widom edge of the density of states. Therefore, this edge scaling domain is obtained by coupling to a constant external matrix $A = 1$, which simply brings the edge of Wigner's semi-circle to the origin.

(ii) As shown in Sect. 2.4, the intersection numbers are described by ribbon graphs, which circulate between the marked point as one single stroke line. For one marked point, the Gaussian means in the $N \to 0$ replica limit, corresponds to one stroke line Feynman diagrams, with trivalent vertices.The computation has been done by this method in [30], reproducing (6.5).

(iii) From the duality relation (4.2), with a unit matrix source $A = 1$ ($a_i = 1$), one obtains a dual model for matrices B. In the large N limit, it reduces to Kontsevich's Airy matrix model. The n-point function $U(\sigma_1, \ldots, \sigma_n)$, with this simple choice of external source A, becomes a generating function of the intersection numbers of curves. For the one point function this yields

$$U(\sigma) = \frac{1}{\sigma} \int \frac{du}{2\pi i} e^{\frac{i}{3}\left[\left(u+\frac{1}{2}\sigma\right)^3 - \left(u-\frac{1}{2}\sigma\right)^3\right]} \tag{6.6}$$

which yields immediately (6.5) if one uses the identification spelled out below in (7.25).

6.3 KdV Hierarchy

Kontsevich's matrix model is

$$\begin{aligned} Z &= \int dM e^{-\frac{i}{6}\mathrm{tr} M^3 - \frac{i}{2}\mathrm{tr} \Lambda^2 M} \\ &= e^{-\mathrm{tr}\frac{\Lambda^3}{3}} \int dM e^{-\frac{i}{6}\mathrm{tr} M^3 - \frac{1}{2}\mathrm{tr} \Lambda M^2} \end{aligned} \tag{6.7}$$

where the matrices M and Λ are $N \times N$ Hermitian matrices, and shift $M \to M - i\Lambda$ is used.

Following Kontsevich [89], we factor Z as

$$Z = \prod_{a,b} (\lambda_a + \lambda_b)^{-\frac{1}{2}} Y \tag{6.8}$$

where λ_a are eigenvalues of Λ.

The free energy $F = \log Z$ in (6.4), which is the generating function of the intersection numbers, has hierarchical structure. Kontsevich has shown that the second derivative of this free energy F with respect to t_0, satisfies the Korteweg de Vries equation (KdV).

$$V = \frac{\partial^2 F}{\partial t_0^2} \tag{6.9}$$

$$\frac{\partial V}{\partial t_1} = V \frac{\partial V}{\partial t_0} + \frac{1}{12} \frac{\partial^3 V}{\partial t_0^3} \qquad \text{(KdV equation)} \tag{6.10}$$

This KdV equation is the first of a hierarchy of differential equations,

$$\frac{\partial V}{\partial t_n} = \frac{\partial R_{n+1}}{\partial t_0} \tag{6.11}$$

with the Gelfand–Dikii differential polynomials R_n.

$$R_1 = V, \quad \frac{\partial R_{n+1}}{\partial t_0} = \frac{1}{2n+1}\left(\frac{\partial V}{\partial t_0} R_n + 2V\frac{\partial R_n}{\partial t_0} + \frac{1}{4}\frac{\partial^3 R_n}{\partial t_0^3}\right). \tag{6.12}$$

This gives for instance,

$$R_2 = \frac{1}{2}V^2 + \frac{1}{12}\frac{\partial^2 V}{\partial t_0^2}. \tag{6.13}$$

Alternatively one can use differential equations (or Virasoro equations) for Y in (6.7), which becomes $L_m Y = 0$, [43, 70].

$$\begin{aligned}
L_{-1} &= -\frac{1}{2}\frac{\partial}{\partial \tilde{t}_0} + \frac{1}{2}\sum_k (2k+1)\tilde{t}_k \frac{\partial}{\partial \tilde{t}_{k-1}} + \frac{1}{4}\tilde{t}_0^2 \\
L_0 &= -\frac{1}{2}\frac{\partial}{\partial \tilde{t}_1} + \frac{1}{2}\sum_k (2k+1)\tilde{t}_k \frac{\partial}{\partial \tilde{t}_k} + \frac{1}{16} \\
L_n &= -\frac{1}{2}\frac{\partial}{\partial \tilde{t}_{n+1}} + \frac{1}{2}\sum_k (2k+1)\tilde{t}_k \frac{\partial}{\partial \tilde{t}_{k+n}} + \frac{1}{4}\sum_k \frac{\partial^2}{\partial \tilde{t}_{k-1}\partial \tilde{t}_{n-k}}.
\end{aligned} \tag{6.14}$$

and the intersection numbers can be computed from the differential equations (6.11) or (6.14). The expansion of Y is obtained from (6.14), by the relation of $\tilde{t}_n$ and t_n as $t_n = (2n+1)!!\tilde{t}_n$,

$$Y = 1 + \left(\frac{1}{6}t_0^3 + \frac{1}{24}t_1\right) + \left(\frac{25}{144}t_0^3 t_1 + \frac{1}{24}t_0 t_2 + \frac{25}{1152}t_1^2 + \frac{1}{72}t_0^6\right) + \cdots. \tag{6.15}$$

and free energy $F = \log Y$ becomes

$$\begin{aligned}
F &= \frac{1}{6}t_0^3 + \frac{1}{24}t_1 + \frac{1}{6}t_0^3 t_1 + \frac{1}{24}t_0 t_2 + \frac{1}{48}t_1^2 \\
&+ \frac{1}{1152}t_4 + \frac{1}{6}t_0^3 t_1^2 + \frac{1}{24}t_0^4 t_2 + \frac{1}{72}t_1^3 + \frac{29}{5760}t_2 t_3 + \cdots.
\end{aligned} \tag{6.16}$$

From (6.4), the intersection numbers for $p = 2$, with genus g determined by the relation $3g - 3 + s = \sum_i n_i$, are evaluated as

$$\begin{aligned}
&\langle \tau_0^3 \rangle_{g=0} = 1, \quad \langle \tau_1 \rangle_{g=1} = \langle \tau_1^2 \rangle_{g=1} = \langle \tau_0 \tau_2 \rangle_{g=1} = \frac{1}{24}, \\
&\langle \tau_4 \rangle_{g=2} = \frac{1}{1152}.
\end{aligned} \tag{6.17}$$

Chapter 7
Intersection Numbers of p-Spin Curves

The duality formula presented in Chap. 4 and the explicit results for the n-point functions with an external source, make it possible to compute the intersection numbers of a moduli space of p-spin curves, a generalization of Kontsevich intersection numbers considered in the previous chapter.

7.1 Moduli Space of p-Spin Curves

Let Σ is Riemann surface of genus g with s marked points, $x_1, \ldots, x_s$. Integer p ($p \geq 2$) is introduced, and label each x_i by an integer m_i, ($m_i = 0, 1, \ldots, p-1$). (after the consideration of following conditions, x_i is labeled by an additional non-negative integer n_i). The canonical line bundle K of Σ has degree $2g - 2$. Let $\mathscr{O}(x_i)$ be the line bundle of degree 1, and the sections are functions with a simple pole at x_i. The line bundle $\mathscr{S} = K \otimes_i \mathscr{O}(x_i)^{-m_i}$ has degree $2g - 2 - \sum_i m_i$. If this degree is divisible by p, then $\mathscr{S}$ has p^{th} roots. There exists a line bundle $\mathscr{T}$ such that $\mathscr{T}^{\otimes p} \simeq \mathscr{S}$.

To each marked point one associates now a dimension n_i and a "spin" m_i. The relation giving the dimension of the compactified moduli space $\overline{\mathscr{M}}_{g,s}$ is now written according to Witten [135]

$$3g - 3 + s = \sum_{1}^{s} n_i + D \tag{7.1}$$

with

$$D = (g-1)\left(1 - \frac{2}{p}\right) + \frac{1}{p}\sum_{1}^{s} m_i \tag{7.2}$$

E. Brézin and S. Hikami, *Random Matrix Theory with an External Source*, SpringerBriefs in Mathematical Physics, DOI 10.1007/978-981-10-3316-2_7

The intersection numbers are now labelled by the double indices, dimension n_i and spin m_i such as $\langle\tau_{n_i,m_i}\rangle_g$ for genus g.

In Kontsevich spinless case the m_i are all equal to zero, and $p = 2$ so that D vanishes. As we will see in a matrix model generalizing Kontsevich Airy model, this double label, dimension and spin of a marked point, corresponds to an expansion in powers of $\mathrm{tr}\frac{1}{\Lambda^{n+\frac{m+1}{p}}}$, with an external source matrix Λ.

The intersection numbers of s-marked points are represented with the first Chern class c_1 and with the top Chern class $c_D(\mathscr{V})$ and a spin value (m $= 0, 1, \ldots, \mathrm{p}-1$), which contribute to the dimensions of the moduli spaces of (7.1). $\mathscr{V}$ is a vector bundle, $\mathscr{V} = H^0(\Sigma, K \otimes \mathscr{T}^{-1}) = H^0(\Sigma, Hom(\mathscr{T}, K))$ [135]. This top Chern class $c_D(\mathscr{V})$ is Euler class, and when $p = -1$, indeed as will be shown later, the intersection numbers becomes Euler characteristics.

$$\langle\prod_{i=1}^{s}\tau_{n_i,m_i}\rangle_g = \frac{1}{p^g}\int_{\overline{\mathscr{M}}_{g,s}^{1/p}}\prod_{i=1}^{s}c_1(\mathscr{L}_i)^{n_i}c_D(\mathscr{V}) \tag{7.3}$$

where $\mathscr{L}_i$ is a cotangent line bundle at a marked point x_i.

The dimensional constraint of (7.1) must hold, otherwise the intersection numbers are vanishing. The generating functions of the intersection numbers of p spin curves as matrix models have been studied in [1, 89, 135]. Witten conjectured that in this p spin curve case, the generating function will satisfy a higher KdV hierarchy. The spinless intersection numbers have been studied within Teichmuller spaces [99], but such studies p spin curves in Teichmuller space have not appeared yet. We mention here the relevant case, in which arithmetic Euler characteristics with character p was studied for Abelian variety. This case has similar automorphism as p-spin curve at least for small genera. Deuring mass formula for genus one is

$$\frac{p-1}{24} = \sum_E \frac{1}{\#\mathrm{Aut}(E)}. \tag{7.4}$$

where the sum is over the number of isomorphism classes of supersingular elliptic curves in characteristics p [128]. For higher genus g, the generalization of above formula is,

$$\sum_A \frac{1}{\#\mathrm{Aut}(A)} = \frac{(-1)^{\frac{g(g+1)}{2}}}{2^g}\prod_{k=1}^{g}\zeta(1-2k)\prod_{k=1}^{g}(p^k + (-1)^k) \tag{7.5}$$

where the sum is over the isomorphism classes of principally polarized abelian varieties, $\zeta(z)$ is Riemann zeta function and $\zeta(1-2k) = (-1)^k\frac{B_k}{2k}$, with Bernoulli number B_k,

$$B_1 = \frac{1}{6}, B_2 = \frac{1}{30}, B_3 = \frac{1}{42}, B_4 = \frac{1}{30},$$

$$B_5 = \frac{5}{66}, B_6 = \frac{691}{2730}, B_7 = \frac{7}{6}, B_8 = \frac{3617}{510}. \tag{7.6}$$

Above formula contains arithmetic Euler characteristics of $\chi(Sp(2g, Z))$, which has been known as [122]

$$\chi(Sp(2g, Z)) = \prod_{k=1}^{g} \zeta(1 - 2k). \tag{7.7}$$

These numbers $\chi(Sp(2g, Z))$ of lower genus ($g = 1, 2, 3$) appear in the coefficients of the intersection numbers of p-spin curves, in Sect. 7.2. The Euler characteristic $\chi(Sp(2g, z))$ is interpreted as a volume of orbifold or Satake's $\mathscr{V}$ manifold [121], which is Gauss-Bonnet volume formula [75].

7.2 Intersection Numbers of Spin Curves

The generalization of Kontsevich Airy matrix model is defined simply as [1, 89],

$$Z_p = \frac{1}{Z_0} \int dB e^{\frac{1}{p+1}\mathrm{tr} B^{p+1} - \mathrm{tr} BA} \tag{7.8}$$

As for the $p = 2$ Airy case, this matrix model may be obtained as resulting from the duality on the expectation values of characteristic polynomials by an appropriate tuning of the external source. For this, one uses an external source A with $(p - 1)$ distinct eigenvalues, each of them being $\frac{N}{p-1}$ times degenerate:
$A = \mathrm{diag}(a_1, \ldots, a_1, \ldots.., a_{p-1}, \ldots, a_{p-1})$.

Theorem 4.1.1 yields the duality formula. The r.h.s. of this duality formula (4.2) reads in this case,

$$\langle \prod_{\alpha=1}^{p-1} \det(a_\alpha - iB)^{\frac{N}{p-1}} \rangle = \langle \exp\left[\frac{N}{p-1} \sum_{\alpha=1}^{p-1} \mathrm{trlog}\left(1 - \frac{iB}{a_\alpha}\right) + \frac{NK}{p-1} \log\left(\prod_{\alpha=1}^{p-1} a_\alpha \right) \right] \rangle \tag{7.9}$$

The left-hand side of (4.2) is $\langle \prod_1^K \det(\lambda_i - M) \rangle$ averaged over $N \times N$ matrices with external source A. The matrices B are thus $K \times K$. The value of K is arbitrary in this analysis, except that it should be large enough to be able to treat as independent the successive powers $\sum_\alpha (1/a_\alpha)^m$.

We now expand the logarithm in powers of B, and chose a source matrix A whose eigenvalues fulfill the following conditions:

$$\sum_{\alpha=1}^{p-1} \frac{1}{a_\alpha^2} = p - 1, \quad \sum_{\alpha=1}^{p-1} \frac{1}{a_\alpha^m} = 0, \quad (m = 3, \ldots, p) \quad \sum_{\alpha=1}^{p-1} \frac{1}{a_\alpha^{p+1}} \neq 0. \tag{7.10}$$

By tuning the external source as (7.10), in the large N limit, the higher order terms except B^{p+1} can be negligible, and the non-Gaussian matrix model is obtained. This provides the matrix model Z_p (7.8).

$$Z_p = \frac{1}{Z_0} \int dB \exp\left[\frac{1}{p+1} \mathrm{tr}(B^{p+1} - \mathrm{tr}(BA)\right] \tag{7.11}$$

In doing so we have rescaled the matrix B by a factor $-i\left[\frac{N}{p-1}\sum \frac{1}{a_\alpha^{p+1}}\right]^{-1/(p+1)}$. It is because we work in the large N-regime, in which the rescaled B is finite, that we can drop the higher powers of B in the expansion of the logarithm. For instance the next one NB^{p+2} after rescaling has a coefficient proportional to $N^{-1/(p+1)}$. This is the scaling at the edge of spectrum. For the $p = 2$ case, it is well known for the edge of semi-circle law.

As a concrete example let us consider the case $p = 3$. The external source $a_1 = 1, a_2 = -1$, with degeneracies $N/2$, is the solution of the previous constraints. In the dual problem we expand up to order B^4, take B of order $N^{-1/4}$, and obtain in the large N limit,

$$\langle \prod_{i=1}^{N} \det(a_i - iB) \rangle = \langle [\det(1 + B^2)]^{\frac{N}{2}} \rangle = \int dB e^{-\frac{N}{4}\mathrm{tr}B^4 - iN\mathrm{tr}BA} \tag{7.12}$$

For generalized Kontsevich matrix model, the intersection numbers are evaluated by several different methods as indicated in the previous section as (i)–(iii).

(i), (ii) the replica method of $\lim_{N\to 0} U(\sigma)$ provides as (2.68),

$$\langle \tau_{\frac{1}{3}(8g-5-j),j} \rangle_g = \frac{1}{(12)^g g!} \frac{\Gamma(\frac{g+1}{3})}{\Gamma(\frac{2-j}{3})} \tag{7.13}$$

where spin value $j = 0$ for $g = 1, 4, 7, 10, \ldots$ and $j = 1$ for $g = 3, 6, 9, \ldots$. For $g = 2, 5, 8, \ldots$, the intersection numbers are zero.

(iii) As alternative approach, instead of replica limit of $U(\sigma)$ ($N \to 0$), the use of formula of $U(\sigma)$ under the external source A is proposed in [32], and developed in [32, 35]. It turns out that $U(\sigma_1, \ldots, \sigma_s)$ is a generating function of the intersection numbers of the moduli space of p-spin curve, when the external source A has a condition which gives the vanishing coefficients of $(u + \frac{\sigma}{2})^m$ (m = 1, ..., p).

Theorem 7.2 ([32])

The following s-point function of Gaussian matrix model with an external source, obtained by the conditions for the external source a_α of (7.10) and in the large N limit, is a generating function of the intersection numbers of the moduli space of p-spin curve, which is defined in (7.3),

$$U(\sigma_1,\ldots,\sigma_s) = \frac{1}{(2i\pi)^s}\int \prod_{i=1}^{s} du_i e^{-C_1\sum_{i=1}^{s}[(u_i+\frac{\sigma_i}{2})^{p+1}-(u_i-\frac{\sigma_i}{2})^{p+1}]}$$
$$\times e^{-C_2\sum_{i=1}^{s}\sigma_i}\det\frac{1}{u_i-u_j+\frac{1}{2}(\sigma_i+\sigma_j)} \tag{7.14}$$

with $C_1 = \frac{NC}{p^2-1}$, $C = \sum_{\alpha=1}^{p-1}\frac{1}{a_\alpha^{p+1}}$, *and* $C_2 = \frac{N}{p-1}\sum_{\alpha=1}^{p-1}\frac{1}{a_\alpha}$. *The numbers* C_1, C *and* C_2 *are obtained from the expansion of (7.9) with the condition (7.10).*

Proof of Theorem 7.2

The proof of Theorem 7.2 is to show that it provides the intersection numbers evaluated from Z_p in (7.8). By the definition of $U(\sigma_1,\ldots,\sigma_s)$ in (2.7), it is a Fourier transform of the resolvents $\langle \mathrm{tr}\frac{1}{\lambda_1-B}\cdots\mathrm{tr}\frac{1}{\lambda_s-B}\rangle$. This resolvent is obtained from the characteristic polynomials by taking replica limit ($k\to 0$ limit). For one point function,

$$\begin{aligned}
U(\sigma) &= \lim_{k\to 0}\frac{1}{k}\int d\lambda e^{\sigma\lambda}\sum_{\alpha}\frac{\partial}{\partial\lambda_a}\langle\prod_{\alpha=1}^{k}\det(\lambda_\alpha-M)\rangle_A|_{\lambda_\alpha=\lambda}\\
&= \lim_{k\to 0}\frac{1}{k}\int d\lambda e^{\sigma\lambda}\sum_{\alpha}\frac{\partial}{\partial\lambda_\alpha}\langle e^{\sum \mathrm{trlog}(\lambda_\alpha-M)}\rangle_A|_{\lambda_\alpha=\lambda}\\
&= \lim_{k\to 0}\frac{1}{k}\int d\lambda e^{\sigma\lambda}k\langle \mathrm{tr}\frac{1}{\lambda-M}e^{k\mathrm{trlog}(\lambda-M)}\rangle_A\\
&= \int d\lambda e^{\sigma\lambda}\langle\mathrm{tr}\frac{1}{\lambda-M}\rangle_A
\end{aligned} \tag{7.15}$$

Noting the duality formula is

$$\frac{1}{Z_0}\langle\prod_{\alpha=1}^{k}\det(\lambda_\alpha-M)\rangle_A = \frac{1}{Z_0'}\langle\prod_{j=1}^{N}\det(a_j-iB)\rangle_\Lambda \tag{7.16}$$

the right hand side of above equation leads to Z_p for generalized Kontsevich model of the intersection numbers of p-spin curves, up to normalization constant,

$$Z_p = \sum_{k_{n,j}}\langle\prod_{n,j}\tau_{n,j}^{k_{n,j}}\rangle\prod_{n,j}\frac{t_{n,j}^{k_{n,j}}}{k_{n,j}!} \tag{7.17}$$

with

$$t_{n,j} = C\mathrm{tr}\frac{1}{\Lambda^{n+\frac{j+1}{p}}} = C\sum_{\alpha=1}^{k}\frac{1}{\lambda_\alpha^{n+\frac{1+j}{p}}} \tag{7.18}$$

where C is a constant. Putting all $\lambda_\alpha = \lambda$, the trace gives a factor k, and $t_{n,j}$ is proportional to k. Then replica limit $k \to 0$ of $U(\sigma)$ of (7.15) selects the single trace, which means the one point marked point ($s = 1$). The Fourier transform in (7.15) gives just the inverse of $\lambda, \sigma = \frac{1}{\lambda}$. Thus, the evaluation of the coefficients of the $\mathrm{tr}\frac{1}{\Lambda^{pn+1+j}}$, is equivalent to the coefficient of the power of σ in $U(\sigma)$. The coefficients are intersection numbers.

This argument is extended to s-marked points case. The coefficients $\prod_{m=1}^{s} \sigma_m^{n_m+(1+j_m)/p}$ of the expansion of $U(\sigma_1, \ldots, \sigma_s)$ corresponds to coefficients of $\prod_m \mathrm{tr}\frac{1}{\Lambda^{n_m+(1+j_m)/p}}$, an intersection number for s-marked points, $\langle \prod_m \tau_{n_m, j_m} \rangle$. □

The expression for one marked point $U(\sigma)$ at the critical point is obtained from (7.14). The determinant in (7.14) becomes $\frac{1}{\sigma}$.

$$U(\sigma) = \frac{1}{\sigma} \int \frac{du}{2i\pi} e^{-\frac{c}{p+1}[(u+\frac{1}{2}\sigma)^{p+1} - (u-\frac{1}{2}\sigma)^{p+1}]} \tag{7.19}$$

with $c = \frac{N}{p-1} \sum \frac{1}{a_\alpha^{p+1}}$. The irrelevant factor $e^{-C_2\sigma}$ is neglected. Expanding the exponent, this one point function is

$$\begin{aligned} U(\sigma) &= \frac{1}{\sigma} \int \frac{du}{2i\pi} \exp[-csu^p] \\ &\times \exp\left[-c\left(\frac{p(p-1)}{3!4}\sigma^3 u^{p-2} + \frac{p(p-1)(p-2)(p-3)}{5!4^2}\sigma^5 u^{p-4} + \cdots\right)\right] \end{aligned} \tag{7.20}$$

This integral yields gamma functions after the replacement $u = (\frac{t}{c\sigma})^{1/p}$,

$$\begin{aligned} U(\sigma) &= \frac{1}{\sigma p \pi} \cdot \frac{1}{(c\sigma)^{1/p}} \int_0^\infty dt t^{\frac{1}{p}-1} e^{-t} \\ &\times e^{-\frac{p(p-1)}{3!4}\sigma^{2+\frac{2}{p}} c^{\frac{2}{p}} t^{1-\frac{2}{p}} - \frac{p(p-1)(p-2)(p-3)}{5!16}\sigma^{4+\frac{4}{p}} c^{\frac{4}{p}} t^{1-\frac{4}{p}} + \cdots} \\ &= \frac{1}{N\sigma\pi} \frac{1}{(c\sigma)^{1/p}} \left[\Gamma\left(1 + \frac{1}{p}\right) - \frac{p-1}{24} y \Gamma\left(1 - \frac{1}{p}\right) \right. \\ &+ \frac{(p-1)(p-3)(1+2p)}{5!4^2 3} y^2 \Gamma\left(1 - \frac{3}{p}\right) \\ &- \frac{(p-5)(p-1)(1+2p)(8p^2-13p-13)}{7!4^3 3^2} y^3 \Gamma\left(1 - \frac{5}{p}\right) \\ &+ (p-7)(p-1)(1+2p)(72p^4 - 298p^3 - 17p^2 + 562p + 281) \\ &\times \left. \frac{1}{9!4^4 15} y^4 \Gamma\left(1 - \frac{7}{p}\right) + \cdots \right] \end{aligned} \tag{7.21}$$

with $y = c^{\frac{2}{p}} \sigma^{2+\frac{2}{p}}$. There appear fraction power of σ due to the double scaling of large N and tuning the external source, although (2.7) has an integer power of σ. The condition of the dimension of moduli space for one marked point $s = 1$ becomes from (7.1)

$$(p+1)(2g-1) = pn + m + 1 \tag{7.22}$$

The intersection numbers $\langle \tau_{n,m} \rangle_g$ for genus g are then determined as

$$U(\sigma) = \sum_g \langle \tau_{n,m} \rangle_g \frac{1}{\pi} \Gamma\left(1 - \frac{1+m}{p}\right) c^{\frac{2g-1}{p}} p^{g-1} \sigma^{(2g-1)(1+\frac{1}{p})} \tag{7.23}$$

For genus one, the intersection number is

$$\langle \tau_{1,0} \rangle_{g=1} = \frac{p-1}{24} \tag{7.24}$$

For more than genus two ($g \geq 2$), the intersection numbers of one marked point become

$$\langle \tau_{n,m} \rangle_{g=2} = \frac{(p-1)(p-3)(1+2p)}{p \cdot 5! 4^2 3} \frac{\Gamma(1-\frac{3}{p})}{\Gamma(1-\frac{1+m}{p})} \tag{7.25}$$

$$\langle \tau_{n,m} \rangle_{g=3} = \frac{(p-5)(p-1)(1+2p)(8p^2-13p-13)}{p^2 \cdot 7! 4^3 3^2} \frac{\Gamma(1-\frac{5}{p})}{\Gamma(1-\frac{1+m}{p})}. \tag{7.26}$$

$$\langle \tau_{n,m} \rangle_{g=4} = \frac{(p-7)(p-1)(1+2p)(72p^4-298p^3-17p^2+562p+281)}{p^3 \cdot 9! 4^4 15} \times \frac{\Gamma(1-\frac{7}{p})}{\Gamma(1-\frac{1+m}{p})} \tag{7.27}$$

$$\langle \tau_{n,m} \rangle_{g=5} = (p-1)(p-3)(p-9)(1+2p)(3+4p)(32p^4-162p^3+p^2 + 326p+163) \frac{1}{p^4} \frac{1}{11! 4^5 3} \frac{\Gamma(1-\frac{9}{p})}{\Gamma(1-\frac{1+m}{p})} \tag{7.28}$$

$$\langle \tau_{n,m} \rangle_{g=6} = (p-1)(p-11)(1+2p)(530688p^8-5830544p^7+16589332p^6 + 8955300p^5-65056373p^4-26944928p^3+85178190p^2 + 80708428p+20177107) \frac{1}{p^5} \frac{1}{13! \cdot 7 \cdot 5 \cdot 4^6 3^3} \frac{\Gamma(1-\frac{11}{p})}{\Gamma(1-\frac{1+m}{p})} \tag{7.29}$$

$$\begin{aligned}\langle\tau_{n,m}\rangle_{g=7} &= (p-1)(p-13)(1+2p)(276480p^{10} - 4162944p^9 + 19373392p^8 \\ &- 15701284p^7 - 85580336p^6 + 90672709p^5 + 223326185p^4 \\ &- 61441286p^3 - 299056874p^2 - 189131035p - 37826207) \\ &\times \frac{1}{p^6}\frac{1}{15!4^7 3^3}\frac{\Gamma(1-\frac{13}{p})}{\Gamma(1-\frac{1+m}{p})}\end{aligned} \tag{7.30}$$

$$\begin{aligned}\langle\tau_{n,m}\rangle_{g=8} &= (p-1)(p-3)(p-5)(p-15)(1+2p)(3+4p)(5+6p) \\ &\times (462976p^8 - 6035600p^7 + 19687956p^6 + 7469268p^5 \\ &- 80449429p^4 - 28891344p^3 + 110585438p^2 \\ &+ 103042188p + 25760547)\frac{1}{p^7}\frac{1}{17! \cdot 5 \cdot 4^8 3^2}\frac{\Gamma(1-\frac{15}{p})}{\Gamma(1-\frac{1+m}{p})}\end{aligned} \tag{7.31}$$

$$\begin{aligned}\langle\tau_{n,m}\rangle_{g=9} &= (p-1)(p-17)(1+2p)(43867 \cdot 4^7 \cdot 90p^{14} \\ &- 444463 \cdot 127 \cdot 3^3 2^{10} p^{13} + 125652557 \cdot 37 \cdot 23 \cdot 2^7 p^{12} \\ &- 6071689831 \cdot 31 \cdot 2^8 p^{11} + 19841 \cdot 11699 \cdot 131 \cdot 7^2 2^3 p^{10} \\ &+ 116212686067 \cdot 643 \cdot 2^2 p^9 - 1409311 \cdot 86627 \cdot 59 \cdot 7 \cdot 3 \cdot 2p^8 \\ &- 1431305011 \cdot 13781 \cdot 17 \cdot 3p^7 + 724878602897547p^6 \\ &+ 65033 \cdot 359 \cdot 137 \cdot 37 \cdot 31^2 \cdot 7 \cdot 3p^5 + 199710238499 \cdot 491 \cdot 3p^4 \\ &- 2544870788486423p^3 - 2461465523248055p^2 \\ &- 940301719307839p - 134328817043977)\frac{1}{p^8}\frac{1}{19!4^9 \cdot 7 \cdot 5 \cdot 3^5} \\ &\times \frac{\Gamma(1-\frac{17}{p})}{\Gamma(1-\frac{1+m}{p})}\end{aligned} \tag{7.32}$$

The integers n and m should satisfy the condition of (7.22) for non-vanishing intersection numbers. The ratio of gamma functions becomes by this condition,

$$\frac{\Gamma(1-\frac{2g-1}{p})}{\Gamma(1-\frac{1+m}{p})} = \frac{\Gamma(1-\frac{2g-1}{p})}{\Gamma(m+2-2g-\frac{2g-1}{p})}. \tag{7.33}$$

Therefore, this number is a rational number. The common denominator of Γ function, $\Gamma(1-\frac{1+m}{p})$, shows that the intersection numbers for $m = p-1$ are always zero, since this gamma function diverges. This shows that the Ramond part, which corresponds to $m = p - 1$, is decoupled from Neveu-Schwarz part ($m = 0, 1, \ldots, p - 2$). The expressions for the intersection numbers of one marked point for arbitrary p up to genus nine ($g = 9$) are verified for the cases of $p = 2$, $p = -1$ and $p \to \infty$, which are known. Also above expressions of the intersection numbers are consistent with the results of $p = 3, 4$ and 5 [91]. The expressions of $p \to \infty$ and $p = -1$ are

$$\lim_{p \to -1} \langle \tau_{n,m} \rangle_g = \zeta(1 - 2g) = (-1)^g \frac{B_g}{2g} \tag{7.34}$$

and

$$\lim_{p \to \infty} \langle \tau_{n,m} \rangle_g = (-1)^g \frac{B_g}{(2g)!(2g)} p^g + O(p^{g-1}) \tag{7.35}$$

where B_g is a Bernoulli number, $B_1 = \frac{1}{6}$, $B_2 = \frac{1}{30}$, $B_3 = \frac{1}{42}, \ldots$. These two expressions will be derived in Sects. 7.5 and 7.7 (Propositions 7.5 and 7.7).

The expression of genus one coincides with the Duering formula for the Euler characteristics $\chi(Sp(2g, Z))$ times 2^{-g} in (7.4). The numerical coefficients of the intersection numbers up to order $g = 4$ coincide with the arithmetic Euler characteristics χ_A; for instance, $\frac{1}{2^4}\zeta(-1)\zeta(-3)\zeta(-5)\zeta(-7) = \frac{1}{9!4^4 15}$, which appears in the coefficient of genus four. For genus five, the coefficient $\frac{1}{11!4^5 3}$ is equal to $\frac{3}{2^5}\zeta(-1)\zeta(-3)\zeta(-5)\zeta(-7)\zeta(-9)$. Thus there appears a difference of a factor 3. For genus $g \geq 5$, the coefficient of the intersection numbers differs from the arithmetic Euler characteristics χ_A. The coincidence of the coefficients of the intersection numbers to arithmetic Euler characteristics χ_g for $g = 2$ and $g = 3$ is understood by considering the difference of dim $M_g = 3g - 3$ and dim $A_g = \frac{1}{2}g(g + 1)$. A_g is Abelian variety of genus g. They become same for $g = 2$ and $g = 3$. ($g = 1$ is exceptional since dim $M_g = 0$. For $g = 4$, these dimensions become different by one, which is related to Schottky problem).

The intersection numbers for small p in (2.15) are explicitly obtained to all order of genus from $U(\sigma)$. This generating function $U(\sigma)$ for $p = 2$, $p = 3$ and $p = 4$ are expressed by simple exponential function and Bessel functions [21, 36]. In the following, we set $c = 1$ in (7.21) for simplicity.

$$U(\sigma) = \frac{1}{2\sqrt{\pi}\sigma^{\frac{3}{2}}} e^{\frac{\sigma^3}{12N^2}} \quad (p = 2) \tag{7.36}$$

$$\begin{aligned} U(\sigma) &= \frac{1}{N\sigma(3N\sigma)^{1/3}} A_i\left(-\frac{N^{2/3}}{4 \cdot 3^{1/3}}\sigma^{8/3}\right) \\ &= \frac{1}{6\sqrt{3}}\left[J_{\frac{1}{3}}\left(\frac{1}{12\sqrt{3}}\sigma^4\right) + J_{-\frac{1}{3}}\left(\frac{1}{12\sqrt{3}}\sigma^4\right)\right] \quad (p = 3) \end{aligned} \tag{7.37}$$

$$
\begin{aligned}
U(\sigma) &= \frac{1}{2\sqrt{8}} e^{\frac{3}{160}\sigma^5} \frac{1}{2\sin(\frac{\pi}{4})} \left[I_{-\frac{1}{4}}\left(\frac{1}{32}\sigma^5\right) + I_{\frac{1}{4}}\left(\frac{1}{32}\sigma^5\right) \right] \\
&= \frac{1}{8} \sum_{m,n=0}^{\infty} \frac{1}{m!n!\Gamma(n+\frac{5}{4})} \left(\frac{3}{160}\right)^m \left(\frac{1}{64}\right)^{2n+\frac{1}{4}} \sigma^{5m+10n+\frac{1}{4}}, \quad (p=4)
\end{aligned} \tag{7.38}
$$

and the intersection numbers, therefore, are written for arbitrary genus g for $p = 2$, and $p = 3$ as

$$
\begin{aligned}
\langle \tau_n \rangle_g &= \frac{1}{(24)^g g!} \qquad (p=2) \\
\langle \tau_{n,m} \rangle_g &= \frac{1}{(12)^g g!} \frac{\Gamma(\frac{g+1}{3})}{\Gamma(\frac{2-m}{3})}, \quad \left(n = \frac{8g-5-m}{3}, m = 0, 1, \quad p = 3 \right).
\end{aligned} \tag{7.39}
$$

For $p = 2$, it reduces to the expression of Kontsevich model in Chap. 6 [89]. For $p = 3$, above result is just expansion of Airy function, as can be seen in explicit integral representation. For generalized Airy functions for $p\rangle 3$, we consider Airy-Hardy integrals. One notes that the Airy-Hardy integrals $Ei_n(x)$ are given by Tchebycheff polynomials [74, 129]. The Tchebycheff polynomials are

$$
T_n(x) = \cos(n\arccos x) \tag{7.40}
$$

with

$$
\begin{aligned}
&T_0(x) = 1, \quad T_1(x) = x, \quad T_2(x) = 2x^2 - 1, \\
&T_3(x) = 4x^3 - 3x, \quad T_4(x) = 8x^4 - 8x^2 + 1.
\end{aligned}
$$

The Airy-Hardy integral (generalized Airy integral) is

$$
Ei_n(x) = \int_0^{\infty} \exp[-T_n(t, x)]dt \tag{7.41}
$$

with a Tchebycheff function $T_n(t, x)$ defined by

$$
T_n(t, x) = 2x^{\frac{n}{2}}(-i)^n T_n\left(\frac{it}{2\sqrt{x}}\right). \tag{7.42}
$$

The generalized Airy integral is expressed as a Bessel function [129].

$$
\begin{aligned}
Ei_{2n}(x) &= \frac{\sqrt{x}}{n} K_{\frac{1}{2n}}(2x^n) \\
Ei_{2n}(-x) &= \pi\sqrt{x}[I_{\frac{1}{2n}}(2x^n) + I_{-\frac{1}{2n}}(2x^n)] \frac{1}{2n\sin(\frac{\pi}{2n})}
\end{aligned} \tag{7.43}
$$

$Ei_n(x)$ satisfies

$$Ei_n''(x) + n^2 x^{n-2} Ei_n(x) = n x^{\frac{(n-3)}{2}} \tag{7.44}$$

The explicit intersection number for $p = 4$ in (7.37) is obtained by the use of Airy-Hardy integral $Ei_4(x)$ in (7.43). For $p > 4$, however the expression as a Airy-Hardy integral is not useful.

In the large p limit, the intersection numbers become

$$\lim_{p\to\infty} \langle \tau_{n,m} \rangle_g = \frac{B_g}{(2g)!(2g)} p^g + O(p^{g-1}) \tag{7.45}$$

where B_g is a Bernoulli number in (9.30). Since the Bernoulli numbers are related to the Riemann zeta function

$$\zeta(2g) = 2^{2g-1} \pi^{2g} \frac{B_g}{(2g)!}, \tag{7.46}$$

this large p limit of the intersection numbers are expressed as

$$\lim_{p\to\infty} \langle \tau_{n,m} \rangle_g = \frac{p^g}{(2\pi)^{2g} g} \zeta(2g) \tag{7.47}$$

The derivation of this result will be given in (7.170) of Sect. 7.7.

For more than one marked point, the formula for $U(\sigma_1, \ldots, \sigma_s)$ for p-spin curves gives the intersection numbers. After rescaling of the parameters, $u_i \to u_i - \frac{\sigma_i}{2}$, $\sigma_i \to \sigma_i/N$, the term of two point function $U(\sigma_1, \sigma_2)$ is expressed as

$$\begin{aligned} &\frac{1}{u_1 - u_2 + \frac{1}{2N}(\sigma_1 + \sigma_2)} \frac{1}{u_1 - u_2 - \frac{1}{2N}(\sigma_1 + \sigma_2)} \\ &= \frac{N}{\sigma_1 + \sigma_2} \int_0^\infty dx e^{x(u_1 - u_2)} \mathrm{sh}\left(\frac{x}{2N}(\sigma_1 + \sigma_2)\right) \end{aligned} \tag{7.48}$$

The two point function $U(\sigma_1, \sigma_2)$, after the external source eigenvalues a_α are tuned to the critical values of p spin curves and by large N limit, becomes

$$\begin{aligned} U(\sigma_1, \sigma_2) = &\frac{2N}{\sigma_1 + \sigma_2} \frac{1}{(2\pi i)^2} \int_0^\infty dx \int du_1 du_2 \mathrm{sh}\left(\frac{1}{2N} x(\sigma_1 + \sigma_2)\right) e^{-(u_1 - u_2)x} \\ \times \exp&\left[-\frac{N}{p^2 - 1} \sum_\alpha \frac{1}{a_\alpha^{p+1}} \left(\sum_i \left(u_i + \frac{1}{2N}\sigma_i\right)^{p+1} - (\sum_i \left(u_i - \frac{1}{2N}\sigma_i\right)^{p+1}\right)\right] \end{aligned} \tag{7.49}$$

Two point function $U(\sigma_1, \sigma_2)$ is expressed in the power series with the intersection numbers $\langle \tau_{n_1,m_1} \tau_{n_2,m_2} \rangle_g$,

$$U(\sigma_1,\sigma_2) = \sum_{n_1,n_2,m_1,m_2} \langle \tau_{n_1,m_1}\tau_{n_2,m_2}\rangle_g p^g \Gamma\left(1-\frac{1+m_1}{p}\right)\Gamma\left(1-\frac{1+m_2}{p}\right) \times \sigma_1^{n_1+(1+m_1)/p}\sigma_2^{n_2+(1+m_2)/p} \tag{7.50}$$

with the condition,

$$2g(p+1) = p(n_1+n_2)+m_1+m_2+2. \tag{7.51}$$

If this condition is not satisfied, the intersection numbers are vanishing. For instance, the intersection numbers of two marked point for $p = 3$ has been evaluated from $U(\sigma_1,\sigma_2)$ [35]. For $p = 3$,

$$U(\sigma_1,\sigma_2) = \frac{2}{(\sigma_1+\sigma_2)(3\sigma_2)^{1/3}}\int_0^\infty dy \ \mathrm{sh}\left(\frac{\sigma_1+\sigma_2}{2}(3\sigma_1)^{\frac{1}{3}}y\right) \times A_i\left(y-\frac{1}{4\cdot 3^{1/3}}\sigma_1^{8/3}\right)A_i\left(-ay-\frac{1}{4\cdot 3^{1/3}}\sigma_2^{8/3}\right) \tag{7.52}$$

where $a = (\frac{\sigma_1}{\sigma_2})^{1/3}$, and Airy function $A_i(y)$ satisfies

$$A_i''(y) = yA_i(y). \tag{7.53}$$

The value of Airy function at origin and its derivative become

$$A_i(0) = \frac{3^{-2/3}}{\Gamma(2/3)} = \frac{1}{2\pi 3^{1/3}}\Gamma\left(\frac{1}{3}\right), \quad A_i'(0) = -\frac{3^{-1/3}}{\Gamma(1/3)} = -\frac{1}{2\pi}\Gamma\left(\frac{2}{3}\right) \tag{7.54}$$

which provide the factor of gamma function in (7.50). By the integration of parts, the intersection numbers of two point are obtained in genus two. The terms of genus two becomes

$$U(\sigma_1,\sigma_2)|_{g=2} = \frac{(A_i(0))^2}{32\cdot 3^{2/3}}(-\sigma_1^{14/3}\sigma_2^{2/3} - \frac{11}{5}\sigma_1^{11/3}\sigma_2^{5/3} - \frac{17}{5}\sigma_1^{8/3} - \frac{11}{5}\sigma_1^{5/3}\sigma_2^{11/3} - \sigma_1^{2/3}\sigma_2^{14/3}). \tag{7.55}$$

From this expression, the intersection numbers of genus two are obtained,

$$\langle\tau_{0,1}\tau_{4,1}\rangle_{g=2} = \frac{1}{864}, \quad \langle\tau_{1,1}\tau_{3,1}\rangle_{g=2} = \frac{11}{4320}, \quad \langle\tau_{2,1}\tau_{2,1}\rangle_{g=2} = \frac{17}{4320}. \tag{7.56}$$

For genus three [35],

$$\langle \tau_{0,0}\tau_{7,1}\rangle_{g=3} = \frac{1}{31104}, \quad \langle \tau_{0,1}\tau_{7,0}\rangle_{g=3} = \frac{1}{15552}$$
$$\langle \tau_{1,0}\tau_{6,1}\rangle_{g=3} = \frac{5}{31104}, \quad \langle \tau_{1,1}\tau_{6,0}\rangle_{g=3} = \frac{19}{77760}$$
$$\langle \tau_{2,0}\tau_{5,1}\rangle_{g=3} = \frac{103}{217728}, \quad \langle \tau_{2,1}\tau_{5,0}\rangle_{g=3} = \frac{47}{77760}$$
$$\langle \tau_{3,0}\tau_{4,1}\rangle_{g=3} = \frac{443}{544320}, \quad \langle \tau_{3,1}\tau_{4,0}\rangle_{g=3} = \frac{67}{77760} \tag{7.57}$$

This results are in agreement with the previous evaluations of recursion relations [86, 92].

For $p\rangle 3$, the same method can be applied, and for instance in the case $p = 4$, the following intersection numbers are obtained.

$$\langle \tau_{0,0}\tau_{2,0}\rangle_{g=1} = \frac{1}{8}, \quad \langle \tau_{1,0}\tau_{1,0}\rangle_{g=1} = \frac{1}{8}, \quad \langle \tau_{0,2}\tau_{1,2}\rangle_{g=1} = \frac{1}{96} \tag{7.58}$$

$$\langle \tau_{0,1}\tau_{4,1}\rangle_{g=2} = \frac{1}{320} \tag{7.59}$$

Further results are obtained up to $p = 7$ in [35].

For general p, from (7.49), the intersection numbers of two point are evaluated in terms of the polynomial of p. For genus one,

$$\langle \tau_{0,0}\tau_{2,0}\rangle_{g=1} = \frac{p-1}{24}, \quad \langle \tau_{0,2}\tau_{1,p-2}\rangle_{g=1} = \frac{p-3}{24p} \tag{7.60}$$

This agrees with the results of special values of $p = 4, 5, 7$ in [53, 91]. For genus two,

$$\langle \tau_{0,0}\tau_{4,2}\rangle_{g=2} = \frac{(p-1)(p-3)(2p+1)}{5760p},$$
$$\langle \tau_{0,1}\tau_{4,1}\rangle_{g=2} = \frac{(p-1)(p-2)(p+2)}{2880p}$$
$$\langle \tau_{0,2}\tau_{4,0}\rangle_{g=2} = \frac{(p-1)(p-3)(2p+11)}{5760p} \tag{7.61}$$

These results are in agreement with the previous results for special values of $p = 4, 5$ [86, 91, 92, 139], obtained by the recursion relations.

The present approach, based on a random matrix theory and use of the generating function of s-point functions, is indeed a very different way of computing intersection numbers.

7.3 s-Point Function

The expression of $U(\sigma_1, \ldots, \sigma_s)$ provides the generating function for the intersection numbers of p spin curves of s marked points due to Theorem 7.2 of (7.14). The three point and four point function are particularly important to find the algebraic structure of the p spin curves.

The generating function, which is free energy F, is defined with the parameter $t_{n,m}$ for the intersection numbers

$$F = \sum_{d_{n,m}} \langle \prod_{n,m} (\tau_{n,m})^{d_{n,m}} \rangle_g \prod_{n,m} \frac{1}{d_{n,m}!} t_{n,m}^{d_{n,m}} \tag{7.62}$$

with $s = \sum_{n,m} d_{n,m}$. This is a generalization of Gaussian case in Sect. 2.2.

In three marked point, $s = 3$, the non-vanishing intersection numbers appears for genus zero. The necessary condition of the dimension of the moduli space, when $g = 0$ and $s = 3$, becomes from (7.1),

$$(p+1) = p \sum_{i=1}^{3} n_i + \sum_{i=1}^{3} m_i + 3 \tag{7.63}$$

Hence

$$n_1 = n_2 = n_3 = 0, \quad \sum_{i=1}^{3} m_i = p - 2. \tag{7.64}$$

This reads to [135]

$$\langle \tau_{0,m_1} \tau_{0,m_2} \tau_{m_3} \rangle_{g=0} = \delta_{m_1+m_2+m_3, p-2} \tag{7.65}$$

The string equation is

$$\frac{\partial F}{\partial t_{0,0}} = \frac{1}{2} \sum_{i,j=0}^{p-2} \eta^{ij} t_{0,i} t_{0,j} + \sum_{n=1}^{\infty} \sum_{m=0}^{p-2} t_{n+1,m} \frac{\partial F}{\partial t_{n,m}} \tag{7.66}$$

with

$$\eta^{ij} = \delta_{i+j, p-2} \tag{7.67}$$

For the convenience, we define $\langle\langle \tau_{n_1,m_1} \cdots \rangle\rangle$ as

$$\langle\langle \tau_{n_1,m_1} \cdots \tau_{n_s,m_s} \rangle\rangle = \frac{\partial}{\partial t_{n_1,m_1}} \cdots \frac{\partial}{\partial t_{n_s,m_s}} F. \tag{7.68}$$

For $t_{n,m} = 0$, it reduces to the intersection numbers $\langle \tau_{n_1,m_1} \cdots \tau_{n_s,m_s} \rangle$.

Proposition 7.3.1
The four point function is expressed as a product of three point functions, which is called as topological recursion equation [135],

$$\begin{aligned}&\langle\langle\tau_{n_1+1,m_1}\tau_{n_2,m_2}\tau_{n_3,m_3}\tau_{n_4,m_4}\rangle\rangle\\&=\sum_{m',m''}\langle\langle\tau_{n_1,m_1}\tau_{n_2,m_2}\tau_{0,m'}\rangle\rangle\eta^{m'm''}\langle\langle\tau_{0,m''}\tau_{n_3,m_3}\tau_{n_4,m_4}\rangle\rangle\end{aligned}\tag{7.69}$$

Proposition 7.3.2
There is a crossing symmetry due to the exchange of the indices since the permutations of 2, 3, 4 yields invariance.

$$\begin{aligned}&\langle\langle\tau_{n_1,m_1}\tau_{n_2,m_2}\tau_{0,m'}\rangle\rangle\eta^{m'm''}\langle\langle\tau_{0,m''}\tau_{n_3,m_3}\tau_{n_4,m_4}\rangle\rangle\\&=\langle\langle\tau_{n_1,m_1}\tau_{n_3,m_3}\tau_{0,m'}\rangle\rangle\eta^{m'm''}\langle\langle\tau_{0,m''}\tau_{n_2,m_2}\tau_{n_4,m_4}\rangle\rangle\end{aligned}\tag{7.70}$$

Corollary 7.3.1
When all $t_{n,m}=0$ for (7.69), it becomes

$$\begin{aligned}\langle\tau_{n_1+1,m_1}\tau_{n_2,m_2}\tau_{n_3,m_3}\tau_{n_4,m_4}\rangle_g&=\sum_{m',m''}\langle\tau_{n_1,m_1}\tau_{n_2,m_2}\tau_{0,m'}\rangle_{g'}\\&\times\eta^{m'm''}\langle\tau_{0,m''}\tau_{n_3,m_3}\tau_{n_4,m_4}\rangle_{g-g'}\end{aligned}\tag{7.71}$$

The dimensional condition of left hand is

$$(p+1)(2g+2)=p(n_1+n_2+n_3+n_4+1)+(m_1+m_2+m_3+m_4)+4\tag{7.72}$$

The dimensional condition for the sum of two terms of right hand is

$$(p+1)(2g+2)=p(n_1+n_2+n_3+n_4+1)+(m_1+m_2+m_3+m_4)+(m'+m'')+6\tag{7.73}$$

Since $m'+m''=p-2$, two dimensional conditions coinside.

Corollary 7.3.2

$$\begin{aligned}&\langle\tau_{n_1,m_1}\tau_{n_2,m_2}\tau_{0,m'}\rangle\eta^{m'm''}\langle\tau_{0,m''}\tau_{n_3,m_3}\tau_{n_4,m_4}\rangle\\&=\langle\tau_{n_1,m_1}\tau_{n_3,m_3}\tau_{0,m'}\rangle\eta^{m'm''}\langle\tau_{0,m''}\tau_{n_2,m_2}\tau_{n_4,m_4}\rangle\end{aligned}\tag{7.74}$$

where $\eta^{m'm''}$ is (7.67).

These Eqs. (7.69), (7.70) are equivalent to Gelfand-Dikii equation in the next section (Proposition 7.4). □

The structure constant C_{ijk} is obtained from F for the genus zero,

$$\langle \tau_{0,m_1}\tau_{0,m_2}\tau_{0,m_3}\rangle_{g=0} = \frac{\partial^3 F}{\partial t_{0,m_1}\partial t_{0,m_2}\partial t_{0,m_3}}|_{t_{n,m}=0} = C_{m_1,m_2,m_3} \tag{7.75}$$

By $\eta^{n,m}$ in (7.67), and with the definition,

$$C_{ij}{}^{k} = \sum_m C_{ijm}\eta^{m,k} \tag{7.76}$$

one finds from the crossing symmetry of (7.74)

$$C_{ij}{}^{k}C_{kml} = C_{im}{}^{k}C_{kjl} \tag{7.77}$$

This is a Witten-Dijkgraaf-Verlinde-Verlinde (WDVV) relation [43, 44, 46, 135].

There are two terms, which come from the connected parts of the longest cycles of the determinant in (7.14),

$$\det(a_{ij})|_{\text{longest}} = a_{12}a_{23}a_{31} + a_{13}a_{21}a_{32} \tag{7.78}$$

with $a_{ij} = (u_i - u_j + \frac{1}{2}(\sigma_i+\sigma_j))^{-1}$. These terms are expressed by the integral form. The term of $a_{12}a_{23}a_{31}$ becomes [32]

$$\begin{aligned}
&\frac{1}{u_1-u_2+\frac{1}{2}(\sigma_1+\sigma_2)}\frac{1}{u_2-u_3+\frac{1}{2}(\sigma_2+\sigma_3)}\frac{1}{u_3-u_1+\frac{1}{2}(\sigma_3+\sigma_1)}\\
&= \frac{2}{\sigma_1+\sigma_2+\sigma_3}\int_0^\infty dx\int_0^\infty dy\ \mathrm{sh}\left(\frac{x}{2}(\sigma_1+\sigma_2+\sigma_3)\right)\\
&\times\left[e^{-\frac{\sigma_2}{2}x-\frac{\sigma_1+\sigma_2}{2}y-(x+y)u_1+yu_2+xu_3} + e^{-\frac{\sigma_2}{2}x-\frac{\sigma_2+\sigma_3}{2}y-xu_1-yu_2+(x+y)u_3}\right]
\end{aligned} \tag{7.79}$$

From (7.14), with the scaling $x \to \sigma_1^{1/p}$, $\sigma_i \to \sigma_i/N$, and with the notation $N' = N(\frac{p-1}{pc})^{1/p}$, the three point function related to the term of $a_{12}a_{23}a_{31}$ is expressed by

$$\begin{aligned}
U_1 &= \frac{2N'}{\sigma_1+\sigma_2+\sigma_3}\left(\frac{1}{\sigma_3}\right)^{\frac{1}{p}}\int_0^\infty dx\int_0^\infty dy\int_{-\infty}^{\infty}\frac{dv_1dv_2dv_3}{(2\pi)^3}\mathrm{sh}\left(\frac{x}{2N'}\sigma_1^{\frac{1}{p}}(\sigma_1+\sigma_2+\sigma_3)\right)\\
&\quad\times e^{-\frac{\sigma_2}{2N'}\sigma_1^{\frac{1}{p}}x-\frac{\sigma_1+\sigma_2}{2N'}\sigma_2^{\frac{1}{p}}y-iv_1(x+(\frac{\sigma_2}{\sigma_1})^{\frac{1}{p}}y)+iyv_2+i(\frac{\sigma_1}{\sigma_3})^{\frac{1}{p}}xv_3}G(v_1)G(v_2)G(v_3)\\
U_2 &= \frac{2N'}{\sigma_1+\sigma_2+\sigma_3}\left(\frac{1}{\sigma_3}\right)^{\frac{1}{p}}\int_0^\infty dx\int_0^\infty dy\int_{-\infty}^{\infty}\frac{dv_1dv_2dv_3}{(2\pi)^3}\mathrm{sh}\left(\frac{x}{2N'}\sigma_1^{\frac{1}{p}}(\sigma_1+\sigma_2+\sigma_3)\right)\\
&\quad\times e^{-\frac{\sigma_2}{2N'}\sigma_1^{\frac{1}{p}}x-\frac{\sigma_2+\sigma_3}{2N'}\sigma_2^{\frac{1}{p}}y-iv_1x-iyv_2+i((\frac{\sigma_1}{\sigma_3})^{\frac{1}{p}}x+(\frac{\sigma_2}{\sigma_3})^{\frac{1}{p}}y)v_3}G(v_1)G(v_2)G(v_3)
\end{aligned} \tag{7.80}$$

where

$$G(v_i) = \exp\left[-\frac{(iv_i)^p}{p} - i^p \sum_{m=1}^{[\frac{p}{2}]} \frac{(-1)^m (p-1)!}{(2m+1)! 2^{2m} (p-2m)! N'^{2m}} \sigma_i^{(2+\frac{2}{p})m} v_i^{p-2m} \right] \tag{7.81}$$

The intersection numbers $\langle \tau_{n_1,m_1} \tau_{n_2,m_2} \tau_{n_3,m_3} \rangle$ is obtained from the coefficients of $\sigma_1^{n_1+\frac{1+m_1}{p}} \sigma_2^{n_2+\frac{1+m_2}{p}} \sigma_3^{n_3+\frac{1+m_3}{p}}$.

For genus zero, the intersection numbers of three marked points are obtained from above expressions. From U_2, the relevant terms of order of genus zero are obtained by the expansions of sh function and exponential functions,

$$\int_0^\infty dx \int_0^\infty dy \int \frac{dv_1 dv_2 dv_3}{(2\pi)^3} \left[\frac{\sigma_1^{\frac{1}{p}}}{\sigma_3^{\frac{1}{p}}} \left(\left(\frac{\sigma_2}{\sigma_3} \right)^{\frac{1}{p}} y v_3 \right)^{q_2} \left(\left(\frac{\sigma_1}{\sigma_3} \right)^{\frac{1}{p}} x v_3 \right)^{q_1} \left(\sigma_3 \sigma_2^{\frac{1}{p}} \frac{1}{2} y \right) \right]$$

$$\times e^{-iv_1 x - iy v_2} G_0(v_1) G_0(v_2) G_0(v_3) \frac{1}{q_1! q_2!} = \sigma_1^{\frac{1+q_1}{p}} \sigma_2^{\frac{1+q_2}{p}} \sigma_3^{\frac{1+(p-2-q_1-q_2)}{p}} \tag{7.82}$$

where $G_0(v) = e^{-\frac{1}{p}(iv)^p}$. Hence, it gives

$$\langle \tau_{0,q_1} \tau_{0,q_2} \tau_{0,p-2-q_1-q_2} \rangle_{g=0} = 1 \tag{7.83}$$

which agrees with (7.65). Hence the generating functions F of the genus zero becomes up to three marked points are obtained as

$$\begin{aligned}
F &= \frac{1}{6} t_{0,0}{}^3, \quad (p=2) \\
F &= \frac{1}{2} t_{0,0}{}^2 t_{0,1}, \quad (p=3) \\
F &= \frac{1}{2} t_{0,0}{}^2 t_{0,2} + \frac{1}{2} t_{0,0} t_{0,1}{}^2, \quad (p=4) \\
F &= \frac{1}{2} t_{0,0}{}^2 t_{0,3} + t_{0,0} t_{0,1} t_{0,2} + \frac{1}{3!} t_{0,1}{}^3, \quad (p=5)
\end{aligned} \tag{7.84}$$

From these expressions, the structure constant C_{ijk} in (7.75) is obtained.

The four point function $U(\sigma_1, \sigma_2, \sigma_3, \sigma_4)$ for p spin curves has an integral expression (7.14). With the same notation as three point function $a_{ij} = (\sigma_i - \sigma_j + \frac{1}{2}(\sigma_i + \sigma_j))^{-1}$, one of the longest cycle of the determinant of $U(\sigma_1, \sigma_2, \sigma_3, \sigma_4)$ is

$$\begin{aligned}
a_{12} a_{23} a_{34} a_{41} &= \frac{1}{\sigma_1 + \sigma_2 + \sigma_3 + \sigma_4} (a_{12} + a_{23})(a_{34} + a_{41}) \\
&\times \left(\frac{1}{u_1 - u_3 + \frac{1}{2}(\sigma_1 + 2\sigma_2 + \sigma_3)} - \frac{1}{u_1 - u_3 - \frac{1}{2}(\sigma_1 + 2\sigma_4 + \sigma_3)} \right) \\
&= -\frac{2}{\sigma_1 + \sigma_2 + \sigma_3 + \sigma_4} \int_0^\infty dx dy dz e^{-x(u_1 - u_3) - \frac{1}{2}(\sigma_2 - \sigma_4) x}
\end{aligned}$$

$$
\times \operatorname{sh}\left(\frac{1}{2}x(\sigma_1+\sigma_2+\sigma_3+\sigma_4)\right)\left[e^{-\frac{1}{2}y(\sigma_1+\sigma_2)-\frac{1}{2}z(\sigma_3+\sigma_4)-y(u_1-u_2)-z(u_3-u_4)}\right.
+ e^{-\frac{1}{2}y(\sigma_1+\sigma_2)-\frac{1}{2}z(\sigma_1+\sigma_4)-y(u_1-u_2)-z(u_4-u_1)} + e^{-\frac{1}{2}y(\sigma_2+\sigma_3)-\frac{1}{2}z(\sigma_3+\sigma_4)-y(u_2-u_3)-z(u_3-u_4)}
\left.+ e^{-\frac{1}{2}y(\sigma_2+\sigma_3)-\frac{1}{2}z(\sigma_1+\sigma_4)-y(u_2-u_3)-z(u_4-u_1)}\right] \tag{7.85}
$$

From these four terms, U_1, U_2, U_3 and U_4 are expressed by the same scaling as three point function, $x \to \sigma_1^{\frac{1}{p}}\frac{1}{N'}$ x, $u_1 \to i\sigma_1^{-\frac{1}{p}}N'v_1$, etc. With the notation $\sigma = \sigma_1+\sigma_2+\sigma_3+\sigma_4$,

$$
U_1 = -\frac{2N'^3}{\sigma}\left(\frac{1}{\sigma_4}\right)^{\frac{1}{p}}\int_0^\infty dxdydz\int\frac{1}{(2\pi)^4}\prod_{i=1}^4 dv_i\operatorname{sh}\left(\frac{x}{2N'}\sigma_1^{\frac{1}{p}}\sigma\right)\prod_{i=1}^4 G(v_i)
\times\exp\left[-\frac{1}{2N'}(\sigma_2-\sigma_4)\sigma_1^{\frac{1}{p}}x-\frac{1}{2N'}(\sigma_1+\sigma_2)\sigma_2^{\frac{1}{p}}y-\frac{1}{2N'}(\sigma_3+\sigma_4)\sigma_3^{\frac{1}{p}}z\right.
\left.-ixv_1-i\left(\frac{\sigma_2}{\sigma_1}\right)^{\frac{1}{p}}yv_1+iyv_2+i\left(\frac{\sigma_1}{\sigma_3}\right)^{\frac{1}{p}}xv_3-izv_3+i\left(\frac{\sigma_3}{\sigma_4}\right)^{\frac{1}{p}}zv_4\right] \tag{7.86}
$$

$$
U_2 = -\frac{2N'^3}{\sigma}\left(\frac{1}{\sigma_4}\right)^{\frac{1}{p}}\int_0^\infty dxdydz\int\frac{1}{(2\pi)^4}\prod_{i=1}^4 dv_i\operatorname{sh}\left(\frac{x}{2N'}\sigma_1^{\frac{1}{p}}\sigma\right)\prod_{i=1}^4 G(v_i)
\times\exp\left[-\frac{1}{2N'}(\sigma_2-\sigma_4)\sigma_1^{\frac{1}{p}}x-\frac{1}{2N'}(\sigma_1+\sigma_2)\sigma_2^{\frac{1}{p}}y-\frac{1}{2N'}(\sigma_1+\sigma_4)\sigma_3^{\frac{1}{p}}z\right.
\left.-ixv_1-i\left(\frac{\sigma_2}{\sigma_1}\right)^{\frac{1}{p}}yv_1+iyv_2+i\left(\frac{\sigma_1}{\sigma_3}\right)^{\frac{1}{p}}xv_3+iz\left(\frac{\sigma_3}{\sigma_4}\right)^{\frac{1}{p}}v_1-i\left(\frac{\sigma_3}{\sigma_4}\right)^{\frac{1}{p}}zv_4\right] \tag{7.87}
$$

$$
U_3 = -\frac{2N'^3}{\sigma}\left(\frac{1}{\sigma_4}\right)^{\frac{1}{p}}\int_0^\infty dxdydz\int\frac{1}{(2\pi)^4}\prod_{i=1}^4 dv_i\operatorname{sh}\left(\frac{x}{2N'}\sigma_1^{\frac{1}{p}}\sigma\right)\prod_{i=1}^4 G(v_i)
\times\exp\left[-\frac{1}{2N'}(\sigma_2-\sigma_4)\sigma_1^{\frac{1}{p}}x-\frac{1}{2N'}(\sigma_2+\sigma_3)\sigma_2^{\frac{1}{p}}y-\frac{1}{2N'}(\sigma_3+\sigma_4)\sigma_3^{\frac{1}{p}}z\right.
\left.-ixv_1+i\left(\frac{\sigma_2}{\sigma_3}\right)^{\frac{1}{p}}yv_3-iyv_2+i\left(\frac{\sigma_1}{\sigma_3}\right)^{\frac{1}{p}}xv_3-izv_3+i\left(\frac{\sigma_3}{\sigma_4}\right)^{\frac{1}{p}}zv_4\right] \tag{7.88}
$$

$$U_4 = -\frac{2N'^3}{\sigma}\left(\frac{1}{\sigma_4}\right)^{\frac{1}{p}} \int_0^\infty dxdydz \int \frac{1}{(2\pi)^4} \prod_{i=1}^4 dv_i \mathrm{sh}\left(\frac{x}{2N'}\sigma_1^{\frac{1}{p}}\sigma\right) \prod_{i=1}^4 G(v_i)$$
$$\times \exp\Big[-\frac{1}{2N'}(\sigma_2-\sigma_4)\sigma_1^{\frac{1}{p}}x - \frac{1}{2N'}(\sigma_2+\sigma_3)\sigma_2^{\frac{1}{p}}y - \frac{1}{2N'}(\sigma_1+\sigma_4)\sigma_3^{\frac{1}{p}}z$$
$$-ixv_1 + i\left(\frac{\sigma_2}{\sigma_3}\right)^{\frac{1}{p}}yv_3 - iyv_2 + i\left(\frac{\sigma_1}{\sigma_3}\right)^{\frac{1}{p}}xv_3 - i\left(\frac{\sigma_3}{\sigma_4}\right)^{\frac{1}{p}}zv_4 + i\left(\frac{\sigma_3}{\sigma_1}\right)^{\frac{1}{p}}zv_1\Big] \tag{7.89}$$

The term $\sigma_1^{2/p}\sigma_2^{2/p}\sigma_3^{(p-1)/p}\sigma_4^{(p-1)/p}$, which yields the intersection numbers, $\langle\tau_{0,1}\tau_{0,1}\tau_{0,p-2}\tau_{p-2}\rangle_0$, is obtained from U_3 for the genus zero, for instance. In the large N' limit (genus zero),

$$U_3 = -\left(\frac{\sigma_1}{\sigma_4}\right)^{\frac{1}{p}} \int_0^\infty dxdydz \int \frac{1}{(2\pi)^4} \prod_{i=1}^4 dv_i x\left(\frac{\sigma_4}{2}\sigma_1^{\frac{1}{p}}x\right)\left(\frac{1}{2}\sigma_3\sigma_2^{\frac{1}{p}}y\right) i\left(\frac{\sigma_2}{\sigma_3}\right)^{\frac{1}{p}} yv_3$$
$$\times \exp\Big[-\frac{i^p}{p}\sum_i v_i^p - ixv_1 - iyv_2 - izv_3\Big]$$
$$\times \exp\Big[i\left(\frac{\sigma_1}{\sigma_3}\right)^{\frac{1}{p}}xv_3 + i\left(\frac{\sigma_2}{\sigma_3}\right)^{\frac{1}{p}}yv_3 + i\left(\frac{\sigma_3}{\sigma_4}\right)^{\frac{1}{p}}zv_4\Big] \tag{7.90}$$

Expanding the last factor, one obtains the term, one obtain

$$\sigma_1^{\frac{2+q_1}{p}}\sigma_2^{\frac{2+q_2}{p}}\sigma_3^{\frac{p-1-q_1-q_2+q_3}{p}}\sigma_4^{\frac{p-1-q_3}{p}}$$

terms. This reads to $\langle\tau_{0,1+q_1}\tau_{0,1+q_2}\tau_{0,p-2-q_1-q_2+q_3}\tau_{0,p-2-q_3}\rangle_{g=0}$.

7.4 Generalized KdV Hierarchy and Gelfand-Dikii Equation

Proposition 7.4 (Witten conjecture [135])

The generating function F of the Intersection numbers of moduli space of p-spin curves satisfy p-th Gelfand-Dikii hierarchy equations;

$$\frac{\partial F}{\partial t_{0,0}\partial t_{n,m}} = -c_{n,m}\mathrm{Res}(Q^{n+\frac{1+m}{p}}) \tag{7.91}$$

with

$$c_{n,m} = \frac{(-1)^n p^{n+1}}{\prod_{l=0}^n (lp+m+1)}, \tag{7.92}$$

and with pseudo differential operator Q.

The differential operator Q is introduced as

$$Q = D^p - \sum_{i=0}^{p-2} u_i(x) D^i \tag{7.93}$$

$$Q^{\frac{1}{p}} = D + \sum_{i\rangle 0} w_i D^{-i} \tag{7.94}$$

with $D = \partial_x$ and $Q^{n+\frac{m}{p}} = (Q^{\frac{1}{p}})^{np+m}$.

The fractional power of the differential operators, pseudo-differential equations, has been investigated by Gelfand and Dikii [64–66]. The Witten conjecture about p-spin curves are studied, and proved by Faber, Shadrin and Zvonkine [58], based upon the Givental's results [67] for Gromov-Witten theory.

For $p = 3$, the generalized KdV equation becomes Boussinesq equation, which is known as the shallow water wave equation (propagation of tsunami). The Boussinesq equation [10] is

$$B_{yy} = a B_{xx} + b(B^2)_{xx} + c B_{xxxx} \tag{7.95}$$

where the suffix means a partial derivative; a, b and c are parameters. In the case of the shallow water equation, y is a time, and x a space coordinate. In terms of $t_{0,0}$ and $t_{0,1}$, the Boussinesq equation becomes

$$\frac{\partial^2 F}{\partial t_{0,1}{}^2} = -\frac{2}{3}\left(\frac{\partial^2 F}{\partial t_{0,0}{}^2}\right)^2 + \frac{\partial^4 F}{\partial t_{0,0}{}^4} \tag{7.96}$$

The pseudo differential operators K is defined as,

$$K = \sum_{i=-\infty}^{n} k_i(x) D^i \tag{7.97}$$

$$D = \frac{i}{\sqrt{p}} \frac{\partial}{\partial x} \tag{7.98}$$

The decomposition of K is

$$K = K_+ + K_-, \quad K_+ = \sum_{i=0}^{\infty} k_i(x) D^i \tag{7.99}$$

and the residue of K is defined as

$$\mathrm{res} K = k_{-1} \tag{7.100}$$

The Gelfand-Dikii equation is

$$i\frac{\partial Q}{\partial t_{n,m}} = [Q_+^{n+\frac{1+m}{p}}, Q]\frac{c_{n,m}}{\sqrt{p}} \tag{7.101}$$

with $[Q_+^{n+\frac{m}{p}}, Q] = -[Q_-^{n+\frac{m}{p}}, Q]$. The Eq. (7.101) becomes a equation of u_i and also for equation of v_i,

$$v_i = -\frac{p}{1+i}\mathrm{res}(Q^{\frac{1+i}{p}}), \quad 0 \le i \le p-2 \tag{7.102}$$

The Gelfand Dikii equation for F now becomes

$$\frac{\partial^2 F}{\partial t_{0,0}\partial t_{0,i}} = v_i \tag{7.103}$$

$$i\frac{\partial Q}{\partial t_{n,m}} = [Q_+^{n+\frac{1+m}{p}}, Q]\frac{c_{n,m}}{\sqrt{p}} \tag{7.104}$$

From these Gelfand-Dikii equation and a string equation, the intersection numbers are obtained. [91, 135]

The Virasoro equations for p-th generalized KdV hierarchy are obtained [45, 62] from $p-1$ matrix model of Douglas [55]. For p-th generalized KdV hierarchy, we need W_n algebra in addition to Virasoro algebra, which will be discussed briefly in chapter eight for open intersection numbers.

7.5 Euler Characteristics and the Negative Value $p = -1$

The negative value $p = -1$ corresponds to the Euler characteristics. This is because the Penner model [114] for Euler characteristics is obtained in this limit in (7.11). From the general expression of the intersection numbers of p-spin curves in (7.24)–(7.27), with $p = -1$, the intersection numbers are

$$\begin{aligned}
\langle\tau\rangle_{g=1} &= -\frac{1}{12}, \quad \langle\tau\rangle_{g=2} = \frac{1}{120} \\
\langle\tau\rangle_{g=3} &= -\frac{1}{252}, \quad \langle\tau\rangle_{g=4} = \frac{1}{240} \\
\langle\tau\rangle_{g=5} &= -\frac{1}{132}, \quad \langle\tau\rangle_{g=6} = \frac{691}{32760} \\
\langle\tau\rangle_{g=7} &= -\frac{7}{84}, \quad \langle\tau\rangle_{g=8} = \frac{3617}{8160} \\
\langle\tau\rangle_{g=9} &= -\frac{43867}{14364}
\end{aligned} \tag{7.105}$$

where we denote $\langle\tau\rangle_g$ as $\langle\tau_{n,m}\rangle_g$ in (7.24)–(7.32), since for $p=-1$ the suffix n, m looses meaning. The index n_i of intersection number $\langle\prod\tau_{n_i,m_i}\rangle$ takes a value of $n_i=0$, since there is no first Chern class $c_1(\mathscr{L}_i)^{n_i}=1$ in (7.3) and only the top Chern class $c_D(\mathscr{V})$ exists, which is the Euler characteristics.

Proposition 7.5
Euler characteristics $\chi(\overline{\mathscr{M}}_{g,1})$ are

$$\chi(\overline{\mathscr{M}}_{g,1})=\zeta(1-2g)=(-1)^g\frac{B_g}{2g} \tag{7.106}$$

where B_g is Bernoulli number in (9.30). Euler characteristics are $\chi(\overline{\mathscr{M}}_{1,1})=-\frac{1}{12}$, $\chi(\overline{\mathscr{M}}_{2,1})=\frac{1}{120}$, $\chi(\overline{\mathscr{M}}_{3,1})=-\frac{1}{252},\ldots$, which coincides with (7.105).

Proof of proposition 7.5
In the limit $p\to -1$,

$$\begin{aligned}U(\sigma)&=\frac{1}{N\sigma}\int\frac{du}{2i\pi}e^{-N\log\frac{u+\frac{1}{2}\sigma}{u-\frac{1}{2}\sigma}}\\&=\frac{1}{N}\int\frac{du}{2i\pi}\left(\frac{u-\frac{1}{2}}{u+\frac{1}{2}}\right)^N\end{aligned} \tag{7.107}$$

By the shift of u as σu, one finds that $U(\sigma)$ is independent of σ. After $u\to\frac{1}{2}u$, by setting

$$\frac{u-1}{u+1}=e^{-y},\quad \left(u=\frac{1+e^{-y}}{1-e^{-y}}\right) \tag{7.108}$$

$U(\sigma)$ becomes

$$U(\sigma)=-\frac{1}{N}\int\frac{dy}{2\pi}\frac{e^{-y}}{(1-e^{-y})^2}e^{-Ny}.$$

With the integration by parts, it becomes ($t=Ny$)

$$U(\sigma)=\frac{1}{N}\int_0^\infty dt\frac{1}{1-e^{-\frac{t}{N}}}e^{-t}=\sum_{n=0}^\infty\frac{B_n}{n}\left(\frac{1}{N}\right)^n(-1)^n \tag{7.109}$$

with $\tilde{B}_0=1$, $\tilde{B}_1=\frac{1}{2}$, $\tilde{B}_2=\frac{1}{6}$, $\tilde{B}_3=0$, $\tilde{B}_4=-\frac{1}{30}$. These another definitions of Bernoulli numbers $\tilde{B}_n$ are used here. $\tilde{B}_{2n}=B_n(-1)^{n+1}$. B_n is defined in (7.6). The Euler characteristics for an orbifold with one marked point is thus given by (7.106), obtained from the limit of p-spin curves with $p\to -1$, and it does agree with the result of [76, 114]. □

Higher correlation functions may be obtained by the dilaton equation (2.47) in a recursive way,

$$\langle(\tau_{1,0})^n\rangle_g=(2g-2+n)\langle(\tau_{1,0})^{n-1}\rangle_g \tag{7.110}$$

with

$$\langle \tau_{1,0} \rangle_g = \chi(\overline{\mathcal{M}}_{g,1}) = \zeta(1-2g) = (-1)^g \frac{B_g}{2g} \tag{7.111}$$

Hence the Euler characteristics of s marked points $\chi(\overline{\mathcal{M}}_{g,s})$ are expressed as

$$\chi(\overline{\mathcal{M}}_{g,s}) = \langle (\tau_{1,0})^s \rangle_g = -\frac{2g-1}{(2g)!}(2g+s-3)! B_g \tag{7.112}$$

This result agrees with the previous results of [76, 114].

The Structure of Polynomial of p in the Intersection Numbers

The expressions (7.24)–(7.32) show an interesting structure. The intersection numbers are factorized for the polynomial of p. The complicated term of p except trivial factors, denoted by $Q(p)$, has a structure as

$$Q(p) = Ap^m - (p+1)B(p) \tag{7.113}$$

where A is a coefficient of highest order term, and $B(p)$ is a polynomial of order $m-2$. For $g = 3$,

$$Q(p) = 8p^2 - 13p - 13 = 8p^2 - 13(p+1). \tag{7.114}$$

For $g = 4$,

$$Q(p) = 72p^4 - 298p^3 - 17p^2 + 562p + 281 = 72p^4 - (p+1)[298p^2 - 281(p+1)]. \tag{7.115}$$

For $g = 5$,

$$Q(p) = 32p^4 - (p+1)[162p^2 - 163(p+1)]. \tag{7.116}$$

These interesting structures of the factor $(p+1)$ for the sub-leading terms in $Q(p)$ remain up to genus nine, and it seems valid to all order in the genus. In the large p limit, the intersection numbers become $\frac{1}{(2g)!2g} B_g$, and the difference with the expression of the Euler characteristics $\zeta(1-2g) = (-1)^g B_g/2g$ comes from the ratio of gamma functions. Therefore, there is an interesting relation between the $p = -1$ Euler characteristics and the limit $p \to \infty$ for intersection numbers. The case $p \to \pm\infty$ case will be discussed in Sect. 7.7. Related to the relation between $p \to \infty$ and $p = -1$ cases, we consider the $(p+1)$ expansion around $p = -1$. For small $p+1$,

$$\begin{aligned} U(\sigma) &= \frac{1}{\sigma} \int \frac{du}{2\pi} e^{\frac{N}{p+1}[(u+\frac{\sigma}{2})^{p+1} - (u-\frac{\sigma}{2})^{p+1}]} \\ &= \frac{1}{\sigma} \int \frac{du}{2\pi} \left(\frac{u+\frac{\sigma}{2}}{u-\frac{\sigma}{2}} \right)^N \left[1 + N\frac{p+1}{2} \log\left(\frac{u+\frac{\sigma}{2}}{u-\frac{\sigma}{2}} \right) \log\left(u^2 - \frac{\sigma^2}{4} \right) \right] + O((p+1)^2) \end{aligned} \tag{7.117}$$

By the change of variable $(u + \frac{\sigma}{2})/(u - \frac{\sigma}{2}) = e^{-y}$, and $y \to y/N$, up to order $(p+1)$, it becomes

$$U(\sigma) = \frac{1}{N}\int \frac{du}{2\pi}\frac{e^{-\frac{y}{N}}}{(1-e^{-\frac{y}{N}})^2}e^{-y}\left[1 - \frac{p+1}{2}y\left(2\log\sigma - \frac{y}{N} - 2\log(1-e^{-\frac{y}{N}})\right)\right] \tag{7.118}$$

The terms of order $(p+1)$ are evaluated by the use of expansion,

$$\log(1-e^{-\frac{y}{N}}) = -\frac{y}{2N} + \log\frac{y}{N} + \sum_{n=1}^{\infty}(-1)^{n-1}\frac{2^{2n-1}B_n}{(2n)!n}\left(\frac{y}{2N}\right)^{2n} \tag{7.119}$$

Thus we find the series of the perturbation of $(p+1)$, and the term of order $(p+1)$ is expressed by product of two Bernouill numbers of (7.119) and (7.109). The coefficient of the highest order of p in $B(p)$ will be evaluated by $1/p$ expansion in Sect. 7.7. Therefore, there exists a relation between $p = -1$ and $p \to \infty$ cases.

Analytic Continuation of Positive p to Negative Value

The analytic continuation of positive $p > 0$ to negative $p < 0$ corresponds to the analytic continuation from $SU(2)_{p-2}/U(1)$ to the $SL(2,R)_{|p|-2}/U(1)$ Wess-Zumino action with level $|p|-2$. The analytic continuation to negative p in the matrix model means

$$Z = \int dBe^{-\frac{1}{|p|-1}\mathrm{tr}B^{-|p|+1}+\mathrm{tr}BA} \tag{7.120}$$

The corresponding $U(\sigma)$ reads

$$U(\sigma) = \frac{1}{N\sigma}\oint\frac{du}{2i\pi}e^{\frac{1}{p+1}[(u+\frac{\sigma}{2})^{p+1}-(u-\frac{\sigma}{2})^{p+1}]} \tag{7.121}$$

with negative value of p. The matrix model of (7.120) is obtained from the duality formula of (4.2) by the tuning a_i. We divide a_i to two groups b_l and c_j, $\{a_i\} = \{b_l, c_j\}$, with $i = 1,\ldots,N$ and $l = 1,\ldots,N_1$, $j = 1,\ldots,N_2$ and $N_1 + N_2 = N$. By formal expansion of large b_l and small c_j, it becomes

$$\begin{aligned}\langle\prod_{i=1}^{N}\det(a_i - iB)\rangle &= \langle\prod_{l=1}^{N_1}\det(b_l - iB)\prod_{j=1}^{N_2}\det(\mathrm{c_j} - \mathrm{iB})\rangle \\ &= \langle e^{\sum_l \mathrm{tr}\log[b_l(1-iB/b_l)]}e^{\sum_j \mathrm{tr}\log[-iB(1+ic_j/B)]}\rangle\end{aligned} \tag{7.122}$$

The following constraint for b_l eliminates the term B^2 in the Gaussian distribution,

$$\sum_{l}^{N_1}\frac{1}{b_l} = 0, \quad \sum_{l}^{N_1}\frac{1}{b_l^2} = 1 \tag{7.123}$$

and constraints for c_j

$$\sum_{j=1}^{N_2} c_j = 0, \quad \sum_{j=1}^{N_2} c_j^2 = 0, \ldots, \sum_{j=1}^{N_2} c_j^m \neq 0 \tag{7.124}$$

provides (7.122) in the scaling limits $N_1, N_2 \to \infty$,

$$Z = \frac{1}{Z_0} \int dB e^{c \,\mathrm{tr} B^{-m} + N_2 \mathrm{tr} \log B + \mathrm{tr} B \Lambda} \tag{7.125}$$

where Z_0 and c are normalization constants, and B is Hermitian matrix. This anti-polynomial matrix model with external source becomes equivalent to (7.120) when there is no logarithmic term. Thus we found for negative p case that the characteristic polynomials of duality formula in (4.2) gives the partition function Z of anti-polynomial matrix model with logarithmic term, which has been discussed in [98]. Also by the tuning of external source b_l and c_j, from the characteristic polynomial, we obtain

$$Z = \int dB e^{\sum_{m_1} c_{m_1} \mathrm{tr} B^{m_1} + \sum_{m_2} c_{m_2} \mathrm{tr} B^{-m_2} + N_2 \mathrm{tr} \log B + \mathrm{tr} B \Lambda} \tag{7.126}$$

and we will discuss in Chap. 8, open intersection numbers for Kontsevich-Penner model for which all coefficients c_{m_1} and c_{m_2} are vanishing except $c_{m_1=3}$. In Chap. 10, Gromov Witten theory of $\mathbf{P}^1$ for stationary sector is shown to be related to above partition function.

7.6 The Negative Value $p = -2$

For $p = -2$, the one-point function reads [30]

$$U(\sigma) = \frac{1}{\sigma} \oint \frac{du}{2i\pi} e^{-\frac{1}{u+\sigma} + \frac{1}{u}} \tag{7.127}$$

The contour in the u-plane is parallel to the imaginary axis through the point $u = -\frac{1}{2}$ at which $\frac{1}{u(u+1)}$ is a maximum in the real direction. The change of variable $u = \frac{1}{2}(-1 + \frac{i}{x\sqrt{\sigma}})$ gives

$$U(\sigma) = -\frac{1}{4\pi\sqrt{\sigma}} \int \frac{dx}{x^2} e^{-\frac{4x^2}{1+\sigma x^2}} \tag{7.128}$$

The expansion forl σ small reduces to Gaussian integrals.

$$U(\sigma)=\frac{1}{2\sqrt{\pi\sigma}}\left(\frac{1}{8}\sigma-\frac{9}{3!2^7}\sigma^2+\frac{9\cdot 25}{5!2^9}\sigma^3-\frac{3^2\cdot 5^2\cdot 7}{2^{18}}\sigma^4+\cdots\right) \tag{7.129}$$

This expansion is indeed the same as the limit when $p\to -2$ from the general p-spin curve case of (7.24)–(7.27).

$$U(\sigma)=\frac{1}{N\sigma^{1+\frac{1}{p}}\pi}\left[\Gamma\left(1+\frac{1}{p}\right)-\frac{p-1}{24}y\Gamma\left(1-\frac{1}{p}\right)+\cdots\right] \tag{7.130}$$

with $y=\sigma^{2+\frac{2}{p}}$. Since the expansion of $U(\sigma)$ in (7.21) for positive p was derived from the Gaussian integrals, the small σ expansion due to this Gaussian integrals can be obtained for negative value of p cases from the analytic continuation from positive to negative values of p.

This result has a relation to the unitary matrix model with an external source, which has been considered before as a simple gauge theory. It is

$$Z=\int dU\exp[\mathrm{tr}(C^\dagger U+CU^\dagger)] \tag{7.131}$$

where U is $N\times N$ unitary matrix ($U^\dagger U=1$), and the external source matrix C is a given fixed complex matrix. Gross and Witten [71] found the third order transition at some critical value g_c, ($C=\frac{N}{g^2}$), and Brézin and Gross [12] found also a third order transition governed by the parameter $\mathrm{tr}(C^\dagger C)^{-\frac{1}{2}}$. This model has a weak coupling expansion, i.e. when the eigenvalues λ_i of $C^\dagger C$ are large, in terms of the parameters

$$t_m=\sum_{i=1}^{N}\frac{1}{\lambda_i^{m-\frac{1}{2}}}=\mathrm{tr}(C^\dagger C)^{m-\frac{1}{2}} \tag{7.132}$$

This unitary matrix model is equivalent to an Hermitian matrix model,

$$Z_B=\int dB e^{\mathrm{tr}\frac{1}{B}+k\mathrm{tr}\log B+\mathrm{tr}B\Lambda} \tag{7.133}$$

with $k=-N$. Here the matrix B is Hermitian $N\times N$, and Λ is an external source. This equivalence has been noted by Mironov et al. [98]. The logarithmic term is similar to the Kontsevich-Penner model, which will be discussed in the next chapter.

Let us return to the duality formula in Theorem 4.1.1,

$$\int_{n\times n}dM\prod_{\alpha=1}^{N}\det(\lambda_\alpha-M)e^{-\frac{1}{2}\mathrm{tr}M^2+\mathrm{tr}MA}=\int_{N\times N}dB\prod_{j=1}^{n}\det(a_j-iB)e^{-\frac{1}{2}\mathrm{tr}B^2+\mathrm{tr}B\Lambda} \tag{7.134}$$

when the external source a_j vanishes. The r.h.s. coincides with Z_B in (7.133) for $n = k$. When we take the zero-rplica limit $n \to 0$, the s-point function $U(\sigma_1, \ldots, \sigma_s)$ gives the Fourier transform of r.h.s. of (7.133) with $s = N$.

The weak coupling expansion of the unitary matrix model has been obtained by [37], and it reads

$$Z_B = \frac{1}{(\prod \lambda_i)^{\frac{k+1}{2}} \Delta(\lambda)} \det \left(\frac{1}{\lambda_i^{\frac{j-1}{2}}} I_{k+j}(2\sqrt{\lambda_i}) \right)_{i,j} \tag{7.135}$$

The large-λ expansion of the Bessel function is given by

$$I_l(2\sqrt{\lambda}) = \frac{e^{2\sqrt{\lambda}}}{\sqrt{4\pi\sqrt{\lambda}}} \left(1 - \frac{l^2 - \frac{1}{4}}{4\sqrt{\lambda}} + \cdots \right). \tag{7.136}$$

Defining

$$Z_0 = \prod_{i\langle j}^{N} \frac{1}{\sqrt{\lambda_i} + \sqrt{\lambda_j}} \prod_{i=1}^{N} \frac{1}{\lambda_i^{\frac{k}{2}}} e^{\sum 2\sqrt{\lambda_i}}, \tag{7.137}$$

the partition function Z_B is expressed as

$$\begin{aligned} Z_B = Z_0 \Bigg[1 &- \frac{(2k+2N)^2 - 1}{16} \sum_{i=1}^{N} \frac{1}{\sqrt{\lambda_i}} + \frac{((2k+2N)^2 - 1)((2k+2N)^2 - 9)}{512} \left(\sum_{i=1}^{N} \frac{1}{\sqrt{\lambda_i}} \right)^2 \\ &+ \frac{((2k+N)^2 - 1)((2k+N)^2 - 9)}{3!4^6} \left(-8 \sum_{i=1}^{N} \frac{1}{(\sqrt{\lambda_i})^3} + ((2k+2N)^2 - 17) \left(\sum_{i=1}^{N} \frac{1}{\sqrt{\lambda_i}} \right)^3 \right) \\ &+ \cdots \Bigg]. \end{aligned} \tag{7.138}$$

When $k = -N$, this expression reduces to the partition function of the unitary matrix model. The above expansion can be compared with the expansion of $U(\sigma)$ for $p = -2$ and it will be discussed later [33].

$$\begin{aligned} U(\sigma) = -\frac{1}{2\sqrt{\pi\sigma}} \Bigg(&-\sigma \frac{1}{8}(4k^2 - 1) + \sigma^2 \frac{1}{3!2^7}(4k^2 - 1)(4k^2 - 9) \\ &- \sigma^3 \frac{1}{5!2^9}(4k^2 - 1)(4k^2 - 9)(4k^2 - 25) + \sigma^4 \frac{1}{21 \cdot 2^{18}} \prod_{j=1}^{4} (4k^2 - (2j-1)^2) \\ &- \sigma^5 \frac{1}{135 \cdot 2^{22}} \prod_{j=1}^{5} ((4k^2 - (2j-1)^2) + O(\sigma^6) \Bigg), \end{aligned} \tag{7.139}$$

which is derived from

$$U(\sigma) = \oint \frac{du}{2\pi i} e^{\frac{1}{\sigma u(u+1)}} \left(\frac{u+1}{u}\right)^k. \tag{7.140}$$

Also from the expression of $U(\sigma)$ for general p of (7.21), this expansion for $k = 0$ is obtained by putting $p = -2$.

Note that the constraints of (7.123) and (7.124) of a_i provide the expression of $U(\sigma)$ as

$$\begin{aligned} U(\sigma) &= \oint \frac{du}{2\pi i} \prod_{i=1}^{N} \frac{u - a_j + \sigma}{u - a_j} e^{\frac{1}{2}\sigma^2 + u\sigma} \\ &= \oint \frac{du}{2\pi i} \prod_{l=1}^{N_1} \frac{1 - (u+\sigma)/b_l}{1 - u/b_l} \prod_{j=1}^{N_2} \left(\frac{u+\sigma}{u}\right) \left(\frac{1 + \frac{c_j}{u+\sigma}}{1 - \frac{c_j}{u}}\right) e^{u\sigma + \frac{1}{2}\sigma^2} \\ &= \int \frac{du}{2\pi i} \left(\frac{u+\sigma}{u}\right)^{N_2} e^{c[\frac{1}{(u+\sigma)^m} - \frac{1}{u^m}]}. \end{aligned} \tag{7.141}$$

This leads to (7.140) for $p = -2$ and gives the general negative values of p expression ($m = |p| - 1$).

This equivalence between the two models for $N = 0$ (replica limit), may be seen also from the Virasoro equations, which were derived by Gross and Newman [70].

$$\frac{\partial^2 Z_U}{\partial C_{ab} \partial C^\dagger_{bc}} = \delta_{ac} Z_U \tag{7.142}$$

Z_U is function of the eigenvalues λ_i of $CC^\dagger$ only, and this equation becomes [12]

$$\frac{\partial^2 Z_U}{\partial \lambda_a^2} + \sum_{a \neq b} \frac{1}{\lambda_a - \lambda_b} \left(\frac{\partial Z_U}{\partial \lambda_a} - \frac{\partial Z_U}{\partial \lambda_b}\right) = \frac{1}{\lambda_a} \left(Z_U - \sum_b \frac{\partial Z_U}{\partial \lambda_b}\right). \tag{7.143}$$

This leads to the differential equations. With $Z_U = Z_0 Y$, it becomes

$$\begin{aligned} -\partial_0 Y &= \sum_{k=0}^{\infty} \left(k + \frac{1}{2}\right) \bar{t}_k \partial_k Y + \frac{1}{16} Y \\ -\partial_n Y &= \sum_{k=0}^{\infty} \left(k + \frac{1}{2}\right) \bar{t}_k \partial_{k+n} Y + \frac{1}{4} \sum_{k=1}^{n} \partial_{k-1} \partial_{n-k} Y, \quad (n \geq 1) \end{aligned} \tag{7.144}$$

with $\bar{t}_k = -\frac{1}{2k+1} \sum_b (\frac{1}{\lambda_b})^{k+\frac{1}{2}}$ and $\partial_k = \frac{\partial}{\partial \bar{t}_k}$. Note that these equations, are consistent with (7.139) for $p = -2$ by appropriate renormalization of t_n, but there is no Virasoro operator L_{-1} in this unitary matrix model. The differential equations of (7.144) can be solved, restricted to single $\bar{t}_n$ terms,

$$\log Y = -\frac{1}{4^2}\bar{t}_0 - \frac{3^2}{4^5}\bar{t}_1 - \frac{5^2 \cdot 3^2}{4^7 \cdot 2}\bar{t}_2 - \frac{5^3 \cdot 3^2 \cdot 7^2}{4^{11}}\bar{t}_3 + \cdots \tag{7.145}$$

and one finds by a multiplication of appropriate normalization factors, it agrees with the single sum terms in $\log Z$ of (7.138). For the comparison of Y with the expansion of $U(\sigma)$ of $p = -2$ in (7.21), both expressions agree with normalization factors, i.e. $(-1)^{k+1}(2k+1)\Gamma(k+\frac{1}{2})/\Gamma(\frac{1}{2})$, which becomes $-1, 3/2, -15/4, 3 \cdot 5 \cdot 7/2^3$, ... This gamma factor is due to the definition of $U(\sigma)$,which has Laplace (Fourier) transform of one particle density of state, and the definition of $\bar{t}_k$. By multiplication of these factor to $U(\sigma)$, we obtain above $\log Y$, since

$$U(\sigma) = \frac{1}{16}\sigma^{1/2} - \frac{3}{2^9}\sigma^{3/2} + \frac{5 \cdot 3}{2^{13}}\sigma^{5/2} - \frac{3 \cdot 5^2 \cdot 7}{2^{19}}\sigma^{7/2} + \cdots. \tag{7.146}$$

The strong coupling expansion for this unitary matrix model follows from the expansion of the Bessel functions for small arguments; it yields

$$Z = C\left(1 + \frac{1}{N}\sum_i \lambda_i + \frac{1}{2(N^2-1)}\left(\sum_i \lambda_i\right)^2 - \frac{1}{2N(N^2-1)}\sum_i \lambda_i^2 + O(\lambda^3)\right) \tag{7.147}$$

where C is a constant.

The equivalent one point function from the p-spin curve correlation function at $p = -2$, after the shift $u \to (u-1)/2$, is

$$\begin{aligned} U(\sigma) &= \frac{1}{2}\oint \frac{du}{2i\pi} e^{\frac{4}{\sigma(u^2-1)}}\left(\frac{u+1}{u-1}\right)^k \\ &= \frac{1}{2}\sum_{m=1}^{\infty} \frac{4^m}{m!\sigma^m}\oint \frac{du}{2\pi i}\frac{1}{(u^2-1)^m}\left(\frac{u+1}{u-1}\right)^k \end{aligned} \tag{7.148}$$

where $k = -N$. This contour integral is around the origin, but since the integral falls off at infinity, it is replaced by the contour over the discontinuities across the cuts which run from $(1, \infty)$ and $(-\infty, -1)$. This contour integral becomes

$$\oint \frac{du}{2i\pi} e^{\frac{4}{\sigma(u^2-1)}}\left(\frac{u+1}{u-1}\right)^k = -\frac{2}{\pi}\sin \pi k \int_1^{\infty} dx \frac{(x+1)^{k-m}}{(x-1)^{k+m}} \tag{7.149}$$

which vanishes when k is integer. With an Euler beta function, one finds

$$\int_1^{\infty} dx \frac{(x+1)^{k-m}}{(x-1)^{k+m}} = 2^{1-2m}(2m-2)!\frac{\Gamma(-k-m+1)}{\Gamma(-k+m)}. \tag{7.150}$$

With the expression

$$\frac{\Gamma(-k-m+1)}{\Gamma(-k+m)} = \frac{1}{k}\prod_{l=1}^{m-1}\frac{1}{k^2-l^2} \tag{7.151}$$

up to an additional constant and to an overall factor, one finds

$$U(\sigma) = \sum_{m=1}^{\infty}\frac{(2m-2)!}{m!\sigma^m}\frac{1}{N}\prod_{l=1}^{m-1}\frac{1}{N^2-l^2} \tag{7.152}$$

where $k = -N$ is inserted. The expansion for large σ corresponds to the strong coupling of (7.147) since $\frac{1}{\sigma^m} = \sum_i \lambda_i^m$. Therefore the strong coupling expansion of the unitary matrix model is indeed recovered from the p-spin curve result with $p = -2$ [33]. The poles in this strong coupling expansion is an anomaly noted by de Wit-'t Hooft, but not a real divergence [49, 120].

7.7 Gauged WZW Model of $SU(2)_k/U(1)$, $SL(2,R)_k/U(1)$ and Black Hole

The Wess-Zumino-Witten (WZW) model [131, 133] of a conformal field theory has interesting relation to random matrix theory with an external source. The random matrix theory with external source provides p-th degenerate singularities, as shown in previous sections, and and their singularities are equivalent to the singularity of $SU(2)_k/U(1)$ Wess-Zumino -Witten model, or conformal Wess-Zumino algebra, where the level k is related to p as $p = k+2$. The matrix model can be interpreted as conformal field theory [90] without external source. The relation of Wess-Zumino-Witten model for $SU(2)_k/U(1)$ coset space to p-th spin curves of Riemann surface, generated by random matrix theory with external source, is more intriguing. This relation can be seen in the Virasoro algebra of p-th generalized KdV hierarchy. In the following, we show the correlation function of Wess-Zumino-Witten model for $SU(2)_k/U(1)$ case, which is related to n-point correlation function of $U(\sigma_1,\ldots,\sigma_n)$ for p-th spin curves.

The Wess-Zumino-Witten model is described by the two-dimensional non-linear σ model with a metric ρ^{ij} plus an additional $\Gamma(\hat{g})$ term [88, 116, 133],

$$I(\hat{g}) = -\frac{1}{8\pi}\int_{\Sigma} d^2\sigma\sqrt{\rho}\rho^{ij}\mathrm{tr}(\hat{g}^{-1}\partial_i\hat{g}\cdot\hat{g}^{-1}\partial_j\hat{g}) - i\Gamma \tag{7.153}$$

where $\hat{g}$ runs over a compact Lie group. When Σ is a Riemann surface without boundary, the $\Gamma(\hat{g})$ reads

$$\Gamma(\hat{g}) = \frac{1}{12\pi}\int_B d^3\sigma\epsilon^{ijk}\mathrm{tr}(\hat{g}^{-1}\partial_i\hat{g}\cdot\hat{g}^{-1}\partial_j\hat{g}\cdot\hat{g}^{-1}\partial_k\hat{g})\cdot \tag{7.154}$$

where $\partial B = \Sigma$. The full action is $I(\hat{g})$ multiplied by an integer k, which becomes the number of level.

$$S = kI(\hat{g}). \tag{7.155}$$

The current algebra of Wess-Zumino-Witten model has been clarified by [88]. The extension to cyclic discrete group Z_N, including two dimensional Ising model (N = 2) and three state Potts model (N = 3), has been analyzed by parafermion current algebra [138].

For the gauged WZW model of coset space $SU(2)/U(1)$, the coset of the Lie group for $\hat{g}$ is made of 2×2 matrices, and the correlation functions are [136]

$$\langle U_{n_1}(z_1) \cdots U_{n_s}(z_s) \rangle = \langle g_{11}(z_1)^{n_1} \cdots g_{11}(z_s)^{n_s} \rangle \tag{7.156}$$

where $U_n(z) = g_{11}^n = e^{in\phi(z)}$, and g_{11} is a (1,1) component of a coset of the Lie group $\hat{g}$,

$$\hat{g} = \begin{pmatrix} e^{i\phi} & 0 \\ 0 & e^{-i\phi} \end{pmatrix}. \tag{7.157}$$

The three point function is given by the so called $\phi - a$ system, a being a connection,

$$\langle e^{in_1\phi(z_1)} e^{in_2\phi(z_2)} e^{in_3\phi(z_3)} \rangle = \int D\phi Da e^{\frac{i}{2\pi}\int_\Sigma \phi\cdot((k+2)f+\frac{R}{2})} \cdot e^{in_1\phi(z_1)} e^{in_2\phi(z_2)} e^{in_3\phi(z_3)} \tag{7.158}$$

where f and R are curvature of $U(1)$ field and Ricci scalar of Σ. This is non-vanishing iff $n_1+n_2+n_3 = k$. This result is well known as chiral algebra of $\mathcal{N} = 2$ minimal model at level $k = p - 2$.

$$\langle U_{n_1}(z_1) U_{n_2}(z_2) U_{n_3}(z_3) \rangle = \delta_{n_1+n_2+n_3,k}. \tag{7.159}$$

This is a chiral ring for genus zero. We have indeed found in (7.83),

$$\langle \tau_{0,n_1} \tau_{0,n_2} \tau_{0,n_3} \rangle_{g=0} = \delta_{n_1+n_2+n_3,k}. \tag{7.160}$$

For genus g, the gravitational operator $\tau_r(U_n)$ can be introduced as r-th gravitational descendant of primary field U_n, and their correlation function is expressed by the formula of the intersection numbers of (7.3).

These considerations exhibit the relation between the parameter p of the p-spin curves and the level k of $SU(2)_k/U(1)$ coset space as

$$p = k + 2. \tag{7.161}$$

This shows that gauged WZW conformal field theory for $SU(2)_k/U(1)$ is equivalent to the present moduli space of p-spin curves. This WZW model has a $\mathcal{N} = 2$ supersymmetry, and the central charge c becomes

$$c = 2 - \frac{6}{p} = 2 - \frac{6}{k+2}. \tag{7.162}$$

The central charge c of the $\mathcal{N} = 2$ supersymmetric minimal model is $3k/(k+2)$ and the central charge of gauged WZW model is (-1) from the above value due to $U(1)$ gauge field. Thus the intersection numbers of the moduli space of p-spin curves describes the correlation functions of the gauged WZW model with level $k = p - 2$. There are interesting relations of p spin curve to parafermion theory, which deals with the cyclic group Z_{p-2}. The central charge c is same for both theories as (7.162) [138]. When $p = 4$, it gives $c = 1/2$, which is the central charge of two dimensional Ising model. Two dimensional 3-state Potts model, which has central charge $c = 4/5$, is relevant to $p = 5$ spin curve since the central charge in (7.162) coincides with 4/5.

We turn to the non compact case for $p < 0$. At the transition point of the unitary matrix model in Sect. 7.5, some relations to a black hole have been discussed [6, 123]. The two dimensional black hole of the symmetry $SL(2, R)/U(1)$ has been investigated [46, 137]. The two dimensional current algebra of $SL(2, R)/U(1)$ is known to be unitary, and the coset $SL(2, R)/U(1)$ is a modular invariant model. The gauged WZW model of this $SL(2, R)_k/U(1)$ is a realization of the analytic continuation of k, from positive to negative value, $k \to -k$ of $SU(2)_k/U(1)$ WZW model, since $SL(2, R)$ has a signature $(-, +, +)$ [54].

The central charge c for the coset $SL(2, R)_k/U(1)$ with level k is therefore,

$$c = \frac{3k}{k-2} - 1 = 2 + \frac{6}{k-2}. \tag{7.163}$$

Comparing to the relation $c = 2 - 6/p$, we find $|p| = |k| - 2$, where p is negative spin value of p-spin curve. There are two interesting regions: (i) $k \to \infty$, where the σ model couples weakly to the black hole, and (ii) $k = \frac{9}{4}$, $(p = -\frac{1}{4})$, which means that the black hole has a central charge $c = 26$, and it is described by a bosonic string.

The gauged WZW model is described by the choice of g as [137]

$$g = \cosh r + (\sinh r) \begin{pmatrix} \cos\theta & \sin\theta \\ \sin\theta & -\cos\theta \end{pmatrix} \tag{7.164}$$

with $\theta = it$; it has a Lorentz signature,

$$d\sigma^2 = dr^2 - (\tanh^2 r) dt^2 \tag{7.165}$$

and the black hole solution has been obtained.

Let us now consider the limit $p \to -\infty$. Since the analytic continuation from positive p to negative value p is valid, this limit corresponds also to $p \to \infty$.

From the expression of the intersection numbers for p-spin curves, the limit $p \to -\infty$ has been obtained in [35]. Starting form

$$U(\sigma) = \frac{1}{N\sigma}\int \frac{du}{2i\pi}\exp\left[-\left(\left(u+\frac{1}{2}\sigma\right)^{p+1} - \left(u-\frac{1}{2}\sigma\right)^{p+1}\right)\right] \tag{7.166}$$

with $\sigma \to \sigma/p$ and $u^{p+1} = x^2$, one finds in the limit $p \to \infty$,

$$\begin{aligned} U(\sigma) &= \frac{2}{N\sigma}\int \frac{dx}{2i\pi} x^{-1+\frac{2}{p}} e^{-\frac{2c}{p+1}x^2(\mathrm{sh}\frac{\sigma}{2})} \\ &= \frac{2}{N\sigma}\Gamma\left(\frac{2}{p}\right)\left(\frac{2c}{p+1}\mathrm{sh}\frac{\sigma}{2}\right)^{-\frac{1}{p}}. \end{aligned} \tag{7.167}$$

This is expanded as

$$U(\sigma) = \frac{2}{N\sigma}\Gamma\left(\frac{2}{p}\right)\left(\frac{2c}{p+1}\right)^{-\frac{1}{2}}\left(\frac{\sigma}{2}\right)^{-\frac{1}{p}}\left[1 - \frac{1}{p}\log\left(\frac{\mathrm{sh}\left(\frac{\sigma}{2}\right)}{\left(\frac{\sigma}{2}\right)}\right)\right]. \tag{7.168}$$

Expanding now

$$\log\left[\frac{\mathrm{sh}(\frac{\sigma}{2})}{(\frac{\sigma}{2})}\right] = \sum_{n=1}^{\infty}(-1)^{n-1}\frac{B_n\sigma^{2n}}{(2n)!2n} \tag{7.169}$$

and dropping the irrelevant terms, we find

$$U(\sigma) = \left[1 - \frac{1}{p}\sum_{n=1}^{\infty}(-1)^{n-1}\frac{B_n}{(2n)!2n}\sigma^{2n}\right]\frac{p}{N\sigma}\Gamma\left(1+\frac{2}{p}\right). \tag{7.170}$$

This large p limit gives the intersection numbers with one marked point.

Proposition 7.7

The intersection numbers of moduli space of p spin curve of one marked point is given in the large p limit as

$$\lim_{p\to\infty}\langle\tau_{n,m}\rangle_g = (-1)^g\frac{B_g}{(2g)!(2g)}p^g. \tag{7.171}$$

The Bernoulli numbers B_g which appear in this expression, are indeed the same as in the $p \to -1$ limit. The difference of numerical factor comes from the gamma function in (7.25)–(7.32).

From the relation

$$\frac{1}{e^\sigma - 1} + \frac{1}{2} - \frac{1}{\sigma} = -\frac{d}{d\sigma}(\sigma U(\sigma)) = \sum_{n=1}^{\infty}(-1)^{n-1}\frac{B_n}{(2n)!}\sigma^{2n-1} \tag{7.172}$$

and by the expression of digamma function,

$$\frac{d}{dz}\log\Gamma(z) = \log z - \frac{1}{2z} - z\frac{d}{dz}\int d\sigma U(\sigma)e^{-\sigma z}, \tag{7.173}$$

the density of state $\rho(E)$ is expressed in this limit as

$$\rho(E) = \frac{d}{dE}\mathrm{Im}(\log\Gamma(iE)) - \frac{\pi}{2} - \frac{1}{2E}. \tag{7.174}$$

This density of state is related to the density of state of the black hole of $SL(2,R)/U(1)$. [35, 73, 79, 94]. Note that this coincides also with the asymptotic density of the non trivial zeros of the ζ-function on the critical line.

Chapter 8
Open Intersection Numbers

The intersection numbers are defined on the moduli space of Riemann surface with s-marked points and genus g. When Riemann surface is cut and has boundary, the open intersection numbers appear. There appear open strings which touch to the boundary. Recent development shows that the intersection numbers of moduli space is described by Kontsevich–Penner model, which is Kontsevich Airy matrix model with the logarithmic potential (Penner model) [4, 112]. This Kontsevich–Penner model has been studied, and we derived from two-matrix model, which originates from time dependent Gaussian matrix model with an external source [32, 34].

8.1 Two Matrix Model

The time dependent Gaussian matrix model can be introduced from the Hamiltonian,

$$H = \frac{1}{2}\mathrm{tr}\dot{M}^2 + \frac{1}{2}\mathrm{tr}M^2 \tag{8.1}$$

$M(t)$ are time-dependent Hermitian matrices, $\dot{M}$ their time derivatives [19, 32]. This is a matrix generalization of a quantum particle in an harmonic potential. Starting from a path integral formulation of this time dependent matrix model one can reduce it to a two-matrix model corresponding to the boundaries of the paths [19].

Therefore we consider a two-matrix Gaussian model

$$P(M_1, M_2) = \frac{1}{Z}\exp\left(-\frac{1}{2}\mathrm{tr}M_1^2 - \frac{1}{2}\mathrm{tr}M_2^2 - c\mathrm{tr}M_1M_2 + \mathrm{tr}M_1A\right) \tag{8.2}$$

where M_1 and M_2 are $N \times N$ Hermitian matrices; c is a coupling constant, which is related to the total time of the path by $c = e^{-t}$. It vanishes for $t \to \infty$. A is an

E. Brézin and S. Hikami, *Random Matrix Theory with an External Source*, SpringerBriefs in Mathematical Physics, DOI 10.1007/978-981-10-3316-2_8

external source with eigenvalues a_i, and Z a normalization constant. The duality for the two matrix model with an external source has been investigated in [32]. The characteristic polynomial for $k_1 + k_2$ points are defined by

$$F_{k_1,k_2} = \langle \prod_{\alpha=1}^{k_1} \det(\lambda_\alpha - M_1) \prod_{\beta=1}^{k_2} \det(\mu_\beta - M_2) \rangle \tag{8.3}$$

where the average is over the distribution (8.2).

Proposition 8.1

$$F_{k_1,k_2} = \int dB_1 dB_2 dD dD^\dagger \exp\Big[-\frac{N}{2}\mathrm{tr}(B_1^2 + B_2^2 + 2D^\dagger D) - iN\sqrt{1-c^2}\mathrm{tr} B_1 \Lambda_1$$
$$-iN\sqrt{1-c^2}\mathrm{tr} B_2 \Lambda_2 - \sum_{i=1}^{N} \mathrm{trlog}(1 - X_i) \Big] \tag{8.4}$$

with

$$X_i = \begin{pmatrix} i\sqrt{1-c^2}\frac{B_1}{a_i} & \sqrt{c(1-c^2)}\frac{D}{a_i} \\ -\sqrt{c(1-c^2)}\frac{D^\dagger}{ca_i} & -i\sqrt{1-c^2}\frac{B_2}{ca_i} \end{pmatrix} \tag{8.5}$$

where B_1 and B_2 are Hermtian matrices, $k_1 \times k_1$ and $k_2 \times k_2$ respectively. D is a complex $k_1 \times k_2$ rectangular matrix. The eigenvalues of Λ_1 are λ_α, and for Λ_2 they are denoted μ_β, $(\alpha = 1, \ldots, k_1,\ \beta = 1, \ldots, k_2)$. The eigenvalues of the external source A are a_i, $(i = 1, \ldots, N)$.

Proof of Proposition 8.1

The product of characteristic polynomials may be written as an integral over Grassmann variables ψ_i^α and χ_i^β, with $\alpha = 1, \ldots, k_1$ and $\beta = 1, \ldots, k_2$,

$$F_{k_1,k_2} = \Big\langle \int d\bar\chi d\chi d\bar\psi d\psi e^{N[\bar\psi_\alpha(\lambda_\alpha - M_1)\psi_\alpha + \bar\chi_\beta(\mu_\beta - M_2)\chi_\beta]} \Big\rangle \tag{8.6}$$

Then the integrations over M_1 and M_2 lead to four-fermion terms that may be dissolved with the help of three auxiliary matrices: B_1 a $k_1 \times k_1$ Hermitian matrix, B_2 a $k_2 \times k_2$ Hermitian matrix and D a complex $k_1 \times k_2$ rectangular matrix. From the identities

$$e^{-\frac{N}{2(1-c^2)}\bar\psi\psi\bar\psi\psi} = \int dB_1 e^{-\frac{N}{2}\mathrm{tr} B_1^2 + \frac{iN}{\sqrt{1-c^2}}\mathrm{tr} B_1 \bar\psi\psi}$$
$$e^{-\frac{N}{2(1-c^2)}\bar\chi\chi\bar\chi\chi} = \int dB_2 e^{-\frac{N}{2}\mathrm{tr} B_2^2 + \frac{iN}{\sqrt{1-c^2}}\mathrm{tr} B_2 \bar\chi\chi}$$
$$e^{\frac{Nc}{1-c^2}\bar\psi\chi\bar\chi\psi} = \int dD dD^\dagger e^{-N\mathrm{tr} D^\dagger D + N\sqrt{\frac{c}{1-c^2}}\mathrm{tr}(D\bar\psi\chi + D^\dagger\bar\chi\psi)} \tag{8.7}$$

after Gaussian integration over M_1 and M_2, and the shifts $B_1 \to B_1 + i\sqrt{1-c^2}$ $\lambda_{\alpha,\alpha'}\delta_{\alpha,\alpha'}$ and $B_2 \to B_2 + i\sqrt{1-c^2}\mu_{\beta,\beta'}\delta_{\beta,\beta'}$ one recovers the duality (8.4). □

We now expand $\log(1-X_i)$ in powers of X_i, and consider the case all $a_i = a = 1-c^2$. The quadratic term $\mathrm{tr}B_1^2$ cancels with the one coming from from the expansion of $\log(1-X)$. Note that B_2^2 is not cancelled due to the coupling constant c. Given the factor N in the exponent, the edge scaling limit corresponds to $B_1 \sim O(N^{-1/3})$, $B_2 \sim O(N^{-1/2})$, $D \sim O(N^{-1/3})$ in the large N limit, and the term $N\mathrm{tr}(D^\dagger D B_2) \sim O(N^{-1/6})$ become negligible. Then, in the large N limit, we obtain

$$Z = \int dB_1 dD^\dagger dD e^{-i\mathrm{tr}B_1\Lambda_1 + \frac{i}{3}\mathrm{tr}B_1^3 + i\mathrm{tr}DD^\dagger B_1} \tag{8.8}$$

where the decoupled matrix B_2 is integrated out. After the integration of matrices $D^\dagger$ and D, we obtain Kontsevich matrix model with a logarithmic potential, Kontsevich–Penner model, [32]

$$Z = \int dB_1 e^{\frac{i}{3}B_1^3 - k_2 \mathrm{tr}\log B_1 - i\mathrm{tr}B_1\Lambda_1} \tag{8.9}$$

which turns out as the generating function of the open intersection numbers. The matrix B is $k_1 \times k_1$ matrix and if we put k_2 by k, and k_1 by N, we obtain Kontsevich–Penner model.

8.2 The Kontsevich–Penner Model

The moduli space of curves has been studied for a closed Riemann surface. When there is a boundary, such as a puncture disk, the boundary conditions requires a theoretical extension of the intersection number. When the marked points are located on the closed cycle of boundary, the intersection numbers are called as open intersection numbers, and this open intersection numbers leads to the extension of Kontsevich intersection numbers. The algebraic and geometric investigations of the open intersection numbers attract interest. The study has been initiated by Pandharipande, Solomon and Tessler [112], and the related open KdV hierarchy and open Virasoro equation have been discussed [38, 39, 119]. A matrix model has been proposed for the hierarchical structures [3–5]. The proposed matrix model is Kontsevich Airy model with an logarithmic potential, Kontsevich–Penner model, which was discussed in the previous Sect. 8.1.

$$Z = \int dM e^{-\frac{1}{3}\mathrm{tr}M^3 + \mathrm{tr}M\Lambda + k\mathrm{tr}\log M} \tag{8.10}$$

As discussed in the Sect. 7.6, (negative $p = -2$ case), the logarithmic potential term appeared in the unitary matrix model, which was related to a strong coupling expansion [98].

Since this logarithmic term is $\exp(k \mathrm{tr} \log M) = (\det M)^k$, it is an insertion in the partition function Z of $(\det M)^k$. This may be interpreted as an insertion of a boundary or a brane.

Although the open intersection numbers are considered for $k = 1$ in [3], we take a general case with arbitrary k. For this Kontsevich–Penner model, the same duality technique can be applied; again the s-point function $U(\sigma_1, \ldots, \sigma_s)$ is a generating function of the intersection numbers.

For instance the intersection numbers for one marked point and genus 1 becomes

$$\langle \tau_1 \rangle_{g=1} = \frac{1 + 12k^2}{24} \tag{8.11}$$

and the intersection numbers of Kontsevich model, $\langle \tau_n \rangle$, are with integer number n, and integer g, as we have discussed in Chap. 6, but we have new terms for Kontsevich–Penner model, such as

$$\langle \tau_0 \tau_{\frac{1}{2}} \rangle_{g=\frac{1}{2}} = k \tag{8.12}$$

with half integer $1/2$ for n, and with half integer $g = 1/2$. The appearance of half-integer genera, a characteristic feature of open intersection numbers, is similar to the case of non-orientable surfaces which will be discussed in the next chapter.

The generating function of the intersection numbers with one marked point is $U(\sigma)$, due to duality and replica argument of previous sections, with

$$U(\sigma) = \frac{1}{\sigma} \int \frac{du}{2\pi} e^{-\frac{c}{3}\left[\left(u+\frac{\sigma}{2}\right)^3 - \left(u-\frac{\sigma}{2}\right)^3\right] + k\log\left(u+\frac{\sigma}{2}\right) - k\log\left(u-\frac{\sigma}{2}\right)} \tag{8.13}$$

Expanding this for small k, one obtains the intersection numbers for open boundaries. The intersection numbers computed by this method [34, 36] agree with the results obtained by Virasoro equations, which will be discussed in the next section.

8.3 Open Virasoro Equations for the Kontsevich–Penner Model

In this section, we discuss open Virasoro equation in relation with the W_n algebras.

Let us start again with the Kontsevich–Penner model

$$Z = \int dM e^{\mathrm{tr}\left(-\frac{1}{3}M^3 + \Lambda M + k\log M\right)} \tag{8.14}$$

with M Hermitian $N \times N$ matrix.

From the trivial equation of motion,

$$\int dM \frac{\partial}{\partial M_{ba}} e^{\mathrm{tr}(-\frac{1}{3}M^3 + \Lambda M + k\log M)} = 0 \tag{8.15}$$

one obtains third order partial differential equation,

$$\left(-\left(\frac{\partial}{\partial \Lambda} \right)^3_{ab} + \left(\Lambda^T \frac{\partial}{\partial \Lambda} \right)_{ab} + (N + k)\delta_{ab} \right) Z = 0 \tag{8.16}$$

Since Z is a function of the eigenvalues λ_i of Λ, it is necessary to to trade this equation for differential equations in terms of eigenvalues.

The matrix Λ has eigenvalues $\lambda_1, \lambda_2, \ldots$ and corresponding orthonormal eigenfunctions $|\phi_a\rangle$.

$$\Lambda|\phi_a\rangle = \lambda_a|\phi_a\rangle. \tag{8.17}$$

An increment matrix $d\Lambda$ is,

$$(\Lambda + d\Lambda)\Big(|\phi\rangle + |d\phi\rangle \Big) = (\lambda + d\lambda)\Big(|\phi\rangle + |d\phi\rangle \Big) \tag{8.18}$$

and from this equation, we obtain at first order

$$(\Lambda - \lambda_a)|d\phi_a\rangle + (d\Lambda - d\lambda_a)|\phi_a\rangle = 0 \tag{8.19}$$

Multiplying $\langle\phi_a|$ from the left side, it becomes

$$\langle\phi_a|d\Lambda|\phi_a\rangle = d\lambda_a \tag{8.20}$$

In an arbitrary fixed orthonormal basis $|b\rangle$, it becomes

$$d\lambda_a = \langle\phi_a|b\rangle\langle b|d\Lambda|c\rangle\langle c|\phi_a\rangle \tag{8.21}$$

Therefore, we obtain the first formula,

$$\frac{\partial\lambda_a}{\partial\Lambda_{bc}} = \langle\phi_a|b\rangle\langle c|\phi_a\rangle \tag{8.22}$$

Note that $\langle\phi_a|b\rangle = U_{ab}$, where U is a unitary matrix. From (8.19), multiplying by $\langle\phi_b|$ $(b \neq a)$ the left hand side,

$$\langle\phi_b|d\phi_a\rangle = \frac{1}{\lambda_a - \lambda_b}\langle\phi_b|d\Lambda|\phi_a\rangle \tag{8.23}$$

Therefore, we have

$$|d\phi_a\rangle = \sum_{b\neq a} \frac{1}{\lambda_a - \lambda_b} |\phi_b\rangle\langle\phi_b|d\Lambda|\phi_a\rangle \tag{8.24}$$

from which follows the second formula,

$$\frac{\partial\langle b|\phi_a\rangle}{\partial\Lambda_{cd}} = \sum_{f\neq a} \frac{1}{\lambda_a - \lambda_f} \langle b|\phi_f\rangle\langle\phi_f|c\rangle\langle d|\phi_a\rangle \tag{8.25}$$

The conjugate of this formula is

$$\frac{\partial\langle\phi_a|b\rangle}{\partial\Lambda_{dc}} = \sum_{f\neq a} \frac{1}{\lambda_a - \lambda_f} \langle\phi_f|b\rangle\langle c|\phi_f\rangle\langle\phi_a|d\rangle \tag{8.26}$$

By the chain rule, we obtain the first derivative,

$$\begin{aligned}\frac{\partial Z}{\partial\Lambda_{ab}} &= \frac{\partial\lambda_c}{\partial\Lambda_{ab}}\frac{\partial Z}{\partial\lambda_c} \\ &= \langle b|\phi_c\rangle\langle\phi_c|a\rangle\left(\frac{\partial Z}{\partial\lambda_c}\right)\end{aligned} \tag{8.27}$$

The formula for higher derivatives of Z are obtained through a systematic use of (8.22) and (8.25). For the second derivative, one obtains

$$\begin{aligned}\left(\frac{\partial^2}{\partial\Lambda^2}\right)_{ab} Z &= \left(\frac{\partial}{\partial\Lambda}\right)_{ad}\left(\frac{\partial}{\partial\Lambda}\right)_{db} Z \\ &= \frac{\partial}{\partial\Lambda_{ad}}\left(\langle b|\phi_c\rangle\langle\phi_c|d\rangle\left(\frac{\partial Z}{\partial\lambda_c}\right)\right)\end{aligned} \tag{8.28}$$

Noting that

$$\begin{aligned}&\langle\phi_c|d\rangle\left(\frac{\partial Z}{\partial\lambda_c}\right)\frac{\partial}{\partial\Lambda_{ad}}\langle b|\phi_c\rangle \\ &= \langle\phi_c|d\rangle\left(\frac{\partial Z}{\partial\lambda_c}\right)\sum_f \frac{1}{\lambda_c - \lambda_f}\langle b|\phi_f\rangle\langle\phi_f|a\rangle\langle d|\phi_c\rangle \\ &= \sum_d \langle b|\phi_c\rangle\langle\phi_c|a\rangle\left(\frac{\partial Z}{\partial\lambda_d}\right)\frac{1}{\lambda_d - \lambda_c}\end{aligned} \tag{8.29}$$

and

$$
\begin{aligned}
&\langle b|\phi_c\rangle\left(\frac{\partial Z}{\partial\lambda_c}\right)\frac{\partial}{\partial\Lambda_{ad}}\langle\phi_c|d\rangle\\
&=\langle b|\phi_c\rangle\left(\frac{\partial Z}{\partial\lambda_c}\right)\sum_f\frac{1}{\lambda_c-\lambda_f}\langle\phi_f|d\rangle\langle d|\phi_f\rangle\langle\phi_c|a\rangle\\
&=\langle b|\phi_c\rangle\langle\phi_c|a\rangle\left(\frac{\partial Z}{\partial\lambda_c}\right)\sum_d\frac{1}{\lambda_c-\lambda_d}
\end{aligned}
\tag{8.30}
$$

we obtain

$$
\left(\frac{\partial^2}{\partial\Lambda^2}\right)_{ab}Z=\langle b|\phi_c\rangle\left(\frac{\partial^2}{\partial{\lambda_c}^2}+\sum_{d\neq c}\frac{1}{\lambda_c-\lambda_d}\left(\frac{\partial Z}{\partial\lambda_c}-\frac{\partial Z}{\partial\lambda_d}\right)\right)\langle\phi_c|a\rangle \tag{8.31}
$$

The third order differentiation is obtained by repeating the same procedure. Denoting by Γ_c the operator

$$
\Gamma_c=\frac{\partial^2}{\partial{\lambda_c}^2}+\sum_{d\neq c}\frac{1}{\lambda_c-\lambda_d}\left(\frac{\partial}{\partial\lambda_c}-\frac{\partial}{\partial\lambda_d}\right) \tag{8.32}
$$

one obtains

$$
\begin{aligned}
&\left(\frac{\partial^3}{\partial\Lambda^3}\right)_{pb}=\left(\frac{\partial}{\partial\Lambda}\right)_{pa}\left(\frac{\partial^2}{\partial\Lambda^2}\right)_{ab}\\
&=\left(\frac{\partial\lambda_c}{\partial\Lambda_{pa}}\right)\frac{\partial}{\partial\lambda_c}\Big(\langle b|\phi_c\rangle\Gamma_c\langle\phi_c|a\rangle\Big)\\
&=\langle b|\phi_c\rangle\langle\phi_c|p\rangle\langle a|\phi_c\rangle\langle\phi_c|a\rangle\frac{\partial\Gamma_c}{\partial\lambda_c}\\
&\quad+\langle\phi_c|p\rangle\langle a|\phi_c\rangle\Gamma_c\langle\phi_c|a\rangle\frac{\partial\langle b|\phi_c\rangle}{\partial\lambda_c}\\
&\quad+\langle\phi_c|p\rangle\langle a|\phi_c\rangle\langle b|\phi_c\rangle\Gamma_c\frac{\partial\langle\phi_c|a\rangle}{\partial\lambda_c}
\end{aligned}
\tag{8.33}
$$

Therefore, we obtain

$$
\left(\frac{\partial^3}{\partial\Lambda^3}\right)_{ab}=\langle b|\phi_c\rangle\left(\frac{\partial\Gamma_c}{\partial\lambda_c}+\sum_{d\neq c}\frac{1}{\lambda_c-\lambda_d}(\Gamma_c-\Gamma_d)\right)\langle\phi_c|a\rangle \tag{8.34}
$$

Using the identity,

$$
\frac{1}{(\lambda_c-\lambda_d)(\lambda_c-\lambda_e)}+\frac{1}{(\lambda_d-\lambda_c)(\lambda_d-\lambda_e)}+\frac{1}{(\lambda_e-\lambda_c)(\lambda_e-\lambda_d)}=0 \tag{8.35}
$$

we obtain the expression in terms of eigenvalues

$$\left(\frac{\partial^3}{\partial \Lambda^3}\right)_{ab} = \frac{\partial^3}{\partial {\lambda_c}^3}$$
$$+\sum_{d\neq c} \frac{1}{\lambda_c - \lambda_d}\left(\frac{\partial}{\partial \lambda_c} - \frac{\partial}{\partial \lambda_d}\right)\left(2\frac{\partial}{\partial \lambda_c} + \frac{\partial}{\partial \lambda_d}\right) - \sum_{d\neq c} \frac{1}{(\lambda_c - \lambda_d)^2}\left(\frac{\partial}{\partial \lambda_c} - \frac{\partial}{\partial \lambda_d}\right)$$
$$+2\sum_{d\neq e,c}\sum_{e\neq c} \frac{1}{(\lambda_c - \lambda_e)(\lambda_e - \lambda_d)}\left(\frac{\partial}{\partial \lambda_c} - \frac{\partial}{\partial \lambda_e}\right) \tag{8.36}$$

If we write

$$\Gamma_c^{(1)} = \frac{\partial}{\partial \lambda_c} \tag{8.37}$$

we have

$$\left(\frac{\partial^2}{\partial \Lambda^2}\right)_{ab} = \langle b|\phi_c\rangle \Gamma_c^{(2)} \langle \phi_c|a\rangle \tag{8.38}$$

$$\Gamma_c^{(2)} = \frac{\partial}{\partial \lambda_c}\Gamma_c^{(1)} + \sum_d \frac{1}{\lambda_c - \lambda_d}(\Gamma_c^{(1)} - \Gamma_d^{(1)}) \tag{8.39}$$

Repeating this procedure, if one defines

$$\left(\frac{\partial^{p+1}}{\partial \Lambda^{p+1}}\right)_{ab} = \langle b|\phi_c\rangle \Gamma_c^{(p+1)} \langle \phi_c|a\rangle \tag{8.40}$$

one finds

$$\Gamma_c^{(p+1)} = \frac{\partial}{\partial \lambda_c}\Gamma_c^{(p)} + \sum_d \frac{1}{\lambda_c - \lambda_d}(\Gamma_c^{(p)} - \Gamma_d^{(p)}) \tag{8.41}$$

with the initial condition

$$\Gamma_c^{(1)} = \frac{\partial}{\partial \lambda_c} \tag{8.42}$$

Therefore, the third derivative for Z becomes

$$\frac{\partial^3 Z}{\partial {\lambda_c}^3} + \sum_{d\neq c} \frac{1}{\lambda_c - \lambda_d}\left(\frac{\partial}{\partial \lambda_c} - \frac{\partial}{\partial \lambda_d}\right)\left(2\frac{\partial}{\partial \lambda_c} + \frac{\partial}{\partial \lambda_d}\right) Z$$
$$-\sum_{d\neq c} \frac{1}{(\lambda_c - \lambda_d)^2}\left(\frac{\partial}{\partial \lambda_c} - \frac{\partial}{\partial \lambda_d}\right) Z$$
$$+2\sum_{d\neq e,c}\sum_{e\neq c} \frac{1}{(\lambda_c - \lambda_e)(\lambda_e - \lambda_d)}\left(\frac{\partial}{\partial \lambda_c} - \frac{\partial}{\partial \lambda_e}\right) Z$$
$$-\lambda_c \frac{\partial Z}{\partial \lambda_c} - (N+k)Z = 0 \tag{8.43}$$

Finally, if one defines the parameters t_n as

$$t_n = \sum_{i=1}^{N} \frac{1}{\lambda_i^{n+\frac{1}{2}}} \tag{8.44}$$

where n takes both integer and half-integer values ($n = 0, \frac{1}{2}, 1, \frac{3}{2}, \ldots$), one finds differentials equation in terms of the t_n. The appearance of half-integers is characteristic of the Kontsevich–Penner model. The derivatives with respect to λ_j are replaced by

$$\frac{\partial}{\partial \lambda_j} = \sum_n \frac{\partial t_n}{\partial \lambda_j} \frac{\partial}{\partial t_n} = -\sum_n \left(n + \frac{1}{2}\right) \frac{1}{\lambda_j^{n+\frac{3}{2}}} \frac{\partial}{\partial t_n} \tag{8.45}$$

$$\frac{\partial^2}{\partial {\lambda_j}^2} = \sum_n \frac{\left(n+\frac{1}{2}\right)\left(n+\frac{3}{2}\right)}{\lambda_j^{n+\frac{5}{2}}} \frac{\partial}{\partial t_n} + \sum_n \sum_m \frac{\left(n+\frac{1}{2}\right)\left(m+\frac{1}{2}\right)}{\lambda_j^{m+n+3}} \frac{\partial^2}{\partial t_n \partial t_m} \tag{8.46}$$

$$\frac{\partial^2}{\partial \lambda_1 \partial \lambda_2} = \sum_n \sum_m \frac{\left(n+\frac{1}{2}\right)\left(m+\frac{1}{2}\right)}{\lambda_1^{n+\frac{3}{2}} \lambda_2^{m+\frac{3}{2}}} \frac{\partial^2}{\partial t_n \partial t_m} \tag{8.47}$$

$$\begin{aligned} \frac{\partial^3}{\partial {\lambda_1}^3} = & -\sum_n \frac{\left(n+\frac{1}{2}\right)\left(n+\frac{3}{2}\right)\left(n+\frac{5}{2}\right)}{\lambda_1^{n+\frac{7}{2}}} \frac{\partial}{\partial t_n} \\ & -\sum_n \sum_m \frac{\left(n+\frac{1}{2}\right)\left(m+\frac{1}{2}\right)\left(2n+m+\frac{9}{2}\right)}{\lambda_1^{n+m+4}} \frac{\partial^2}{\partial t_n \partial t_m} \\ & -\sum_n \sum_m \sum_j \frac{\left(n+\frac{1}{2}\right)\left(m+\frac{1}{2}\right)\left(j+\frac{1}{2}\right)}{\lambda_1^{n+m+j+\frac{9}{2}}} \frac{\partial^3}{\partial t_n \partial t_m \partial t_j} \end{aligned} \tag{8.48}$$

where $n, m, j = 0, \frac{1}{2}, 1, \frac{3}{2}, 2, \cdots$.

The zero-th order contribution for large Λ, is obtained from the shift $M \to M + \Lambda^{\frac{1}{2}}$; then keeping only the terms which grow for large Λ one finds

$$\begin{aligned} Z_0 &= \int dM e^{-\mathrm{tr} M^2 \Lambda^{\frac{1}{2}} + \frac{2}{3} \mathrm{tr} \Lambda^{\frac{3}{2}} + \frac{k}{2} \mathrm{tr} \log \Lambda} \\ &= \frac{1}{\prod_{i,j} (\sqrt{\lambda_i} + \sqrt{\lambda_j})^{\frac{1}{2}}} e^{\frac{2}{3} \sum \lambda_i^{\frac{3}{2}}} \prod_i \lambda_i^{\frac{k}{2}} \end{aligned} \tag{8.49}$$

In the limit $\lambda_i \to \infty$, the partition function reduces to Z_0. Therefore, it is convenient to express the partition function Z as

$$Z = Z_0 g(\lambda) \tag{8.50}$$

where g has an expansion in inverse powers of $\sqrt{\lambda}$:

$$g = 1 + O\left(\frac{1}{\lambda^{\frac{3}{2}}}\right) \tag{8.51}$$

As observed in [34], the analysis for small numbers of N, for instance $N = 2$, is useful. With the expansion in the power of $\lambda_1^{-(n+1/2)}$, the equation of (8.43) reduces to the equations of g of order $\lambda^{-(n+1/2)}$. By the use (8.44), we obtain the first equation of order $\lambda^{-1/2}$,

$$\left(-\frac{\partial}{\partial t_0} + \frac{1}{4}t_0^2 - \frac{k}{2}t_{\frac{1}{2}} + \sum_{n=0,\frac{1}{2},1,\ldots}\left(n + \frac{1}{2}\right)t_{n+1}\frac{\partial}{\partial t_n}\right)g = 0 \tag{8.52}$$

Using $\tilde{F} = \ln g$, one obtain the string equation,

$$\frac{\partial \tilde{F}}{\partial t_0} = \frac{1}{4}t_0^2 - \frac{k}{2}t_{\frac{1}{2}} + \sum_{n=0,\frac{1}{2},1,\ldots}\left(n + \frac{1}{2}\right)t_{n+1}\frac{\partial \tilde{F}}{\partial t_n} \tag{8.53}$$

The next order is

$$\begin{aligned}
&\left(-2\frac{\partial}{\partial t_{\frac{1}{2}}} - kt_0 - \frac{k}{4}{t_{\frac{1}{2}}}^2 - \frac{k}{2}t_0t_1 - \frac{1}{4}{t_0}^2t_{\frac{1}{2}} - \frac{1}{16}t_{\frac{3}{2}} - \frac{1}{4}k^2t_{\frac{3}{2}}\right.\\
&- \sum_{n=0,\frac{1}{2},1,\ldots}(2n+1)t_{n+\frac{1}{2}}\frac{\partial}{\partial t_n} + k\sum_{n=0,\frac{1}{2},1,\ldots}\left(n + \frac{1}{2}\right)t_{n+2}\frac{\partial}{\partial t_n}\\
&\left.-\frac{1}{2}\sum_{-i-j+k=-\frac{3}{2}}\left(k + \frac{1}{2}\right)t_it_j\frac{\partial}{\partial t_k} - \frac{1}{2}\sum_{-i+j+k=-\frac{5}{2}}\left(j + \frac{1}{2}\right)\left(k + \frac{1}{2}\right)t_i\frac{\partial^2}{\partial t_j\partial t_k}\right)g = 0
\end{aligned} \tag{8.54}$$

The next order is proportional to ${\lambda_1}^{-\frac{3}{2}}$, and we obtain the dilaton equation,

$$\begin{aligned}
&\left(-3\frac{\partial}{\partial t_1} + \frac{1}{16} + \frac{3}{4}k^2\right.\\
&\left.+ \sum_{n=0,\frac{1}{2},1,\ldots}\left(\frac{1}{2} + n\right)t_n\frac{\partial}{\partial t_n} + \sum_{n=0,\frac{1}{2},1,\ldots}\left(n + \frac{1}{2}\right)t_{n+\frac{3}{2}}\frac{\partial}{\partial t_n}\right)g = 0
\end{aligned} \tag{8.55}$$

These equations determine the free energy $\tilde{F} = \ln g$ up to order $O(\lambda^{-\frac{9}{2}})$,

$$
\begin{aligned}
\tilde{F} &= \frac{1}{12}t_0^3 + \left(\frac{1}{48} + \frac{1}{4}k^2\right) t_1 - \frac{1}{2}kt_0t_{\frac{1}{2}} + \frac{1}{24}t_0^3 t_1 \\
&+ \left(\frac{1}{192} + \frac{1}{16}k^2\right) t_1^2 - \frac{1}{4}kt_0t_{\frac{1}{2}}t_1 - \frac{1}{24}kt_{\frac{1}{2}}^3 + \left(\frac{1}{32} + \frac{3}{8}k^2\right) t_0t_2 - \frac{1}{4}kt_0^2t_{\frac{3}{2}} \\
&+ \frac{1}{4}k^2t_{\frac{1}{2}}t_{\frac{3}{2}} - \frac{1}{6}(k+k^3)t_{\frac{5}{2}} \\
&+ \frac{1}{64}t_0^4t_2 - \frac{1}{6}kt_0^3t_{\frac{5}{2}} + \frac{1}{48}t_0^3t_1^2 + \left(\frac{5}{128} + \frac{15}{32}k^2\right) t_0^2t_3 \\
&- \frac{3}{16}kt_0^2t_{\frac{1}{2}}t_2 - \frac{1}{4}kt_0^2t_1t_{\frac{3}{2}} - \frac{1}{2}(k+k^3)t_0t_{\frac{7}{2}} + \frac{1}{2}k^2t_0t_{\frac{1}{2}}t_{\frac{5}{2}} \\
&+ \left(\frac{1}{32} + \frac{3}{8}k^2\right) t_0t_1t_2 + \frac{1}{4}k^2t_0t_{\frac{3}{2}}^2 - \frac{1}{8}kt_0t_{\frac{1}{2}}^2t_{\frac{3}{2}} - \frac{1}{8}kt_0t_{\frac{1}{2}}t_1^2 \\
&- \left(\frac{15}{64}k + \frac{5}{16}k^3\right) t_{\frac{1}{2}}t_3 + \frac{3}{16}k^2t_{\frac{1}{2}}^2t_2 + \frac{1}{4}k^2t_{\frac{1}{2}}t_1t_{\frac{3}{2}} - \frac{1}{6}(k+k^3)t_1t_{\frac{5}{2}} \\
&+ \left(\frac{1}{576} + \frac{1}{48}k^2\right) t_1^3 - \left(\frac{1}{8}k + \frac{1}{4}k^3\right) t_{\frac{3}{2}}t_2 + \frac{1}{9}\left(\frac{105}{1024} + \frac{735}{128}k^2 + \frac{105}{64}k^4\right) t_4 \\
&- \frac{1}{24}kt_{\frac{1}{2}}^3t_1
\end{aligned} \tag{8.56}
$$

This expansion agrees with the result of [4] by the changes of our parameters to the parameters of A.2 in [4] as $k \to -N$, $t_n \to (2n+1)2^{(2n+1)/3}t_{2n+1}$, $(n = 0, 1/2, 1, 3/2, 2, \ldots)$, due to a different normalization of Kontsevich–Penner model of (8.14). To express these equations in a more compact form, it is convenient to introduce the differential operators $J_n^{(k)}$, obtained as follows [1]. First let us denote by x_n,

$$
x_n = \frac{1}{n}t_{\frac{n-1}{2}} \tag{8.57}
$$

Then the differential operators

$$
J_m{}^{(1)}(x) = \frac{\partial}{\partial x_m} - mx_{-m}, \qquad (m = \ldots, -2, -1, 0, 1, 2, \ldots) \tag{8.58}
$$

and $x_m = 0$ for $x \geq 0$. We now define $J_m^{(2)}$ from $J_m^{(1)}$ as

$$
J_m^{(2)} = \sum_{i+j=m} : J_i^{(1)} J_j^{(1)} : \tag{8.59}
$$

where $: \cdots :$ means a normal ordering, i.e. pulling the differential operator to the right. Then we obtain

$$
J_m{}^{(2)} = \sum_{i+j=m} \frac{\partial^2}{\partial x_i \partial x_j} + 2\sum_{-i+j=m} ix_i \frac{\partial}{\partial x_j} + \sum_{-i-j=m} (ix_i)(jx_j) \tag{8.60}
$$

$$J_m{}^{(3)} = \sum_{i+j+k=m} : J_i^{(1)} J_j^{(1)} J_k^{(1)} :$$
$$= \sum_{i+j+k=m} \frac{\partial^3}{\partial x_i \partial x_j \partial x_k} + 3 \sum_{-i+j+k=m} i x_i \frac{\partial^2}{\partial x_j \partial x_k}$$
$$+ 3 \sum_{-i-j+k=m} (i x_i)(j x_j) \frac{\partial}{\partial x_k} + \sum_{-i-j-k=m} (i x_i)(j x_j)(k x_k) \quad (8.61)$$

where $i, j, k = 1, 2, 3, \ldots$

Returning to the variables t_n

$$x_n = \frac{1}{n} t_{\frac{n-1}{2}}, \quad (8.62)$$

we find

$$J_{-4}^{(2)} = 2t_0 t_1 + t_{\frac{1}{2}}^2 + 4 \sum_{n=0,\frac{1}{2},1,\ldots} \left(n + \frac{1}{2}\right) t_{n+2} \frac{\partial}{\partial t_n} \quad (8.63)$$

$$J_{-2}^{(2)} = t_0^2 + 2 \sum_{n=0,\frac{1}{2},1,\ldots} (2n+1) t_{n+1} \frac{\partial}{\partial t_n} \quad (8.64)$$

$$J_{-1}^{(2)} = 4 \sum_{n=0,\frac{1}{2},1,\ldots} \left(n + \frac{1}{2}\right) t_{n+\frac{1}{2}} \frac{\partial}{\partial t_n} \quad (8.65)$$

$$J_0^{(2)} = 4 \sum_{n=0,\frac{1}{2},1,\ldots} \left(n + \frac{1}{2}\right) t_n \frac{\partial}{\partial t_n} \quad (8.66)$$

From (8.61), we have

$$J_4^{(3)} = 3t_0^2 t_{\frac{1}{2}} + 3 \sum_{-i+j+k=-\frac{5}{2}} (2j+1)(2k+1) t_i \frac{\partial^2}{\partial t_j \partial t_k}$$
$$+ 3 \sum_{-i-j+k=-\frac{3}{2}} \left(k + \frac{1}{2}\right) t_i t_j \frac{\partial}{\partial t_k} \quad (8.67)$$

Then, the first Virasoro constraint is

$$\left(-\frac{\partial}{\partial t_0} + \frac{1}{4} J_{-2}^{(2)} - \frac{k}{2} t_{\frac{1}{2}}\right) g = 0 \quad (8.68)$$

The second equation becomes

$$\left(-2\frac{\partial}{\partial t_{\frac{1}{2}}} - kt_0 - \frac{1}{16}t_{\frac{3}{2}} - \frac{k^2}{4}t_{\frac{3}{2}} - \frac{1}{12}J^{(3)}_{-4} + \frac{k}{4}J^{(2)}_{-4} - \frac{1}{2}J^{(2)}_{-1} \right) g = 0 \tag{8.69}$$

The third equation is expressed by

$$\left(-3\frac{\partial}{\partial t_1} - \frac{1}{16} - \frac{3}{4}k^2 + kt_0 t_{\frac{1}{2}} - \frac{1}{4}J_0{}^{(2)} - \frac{1}{4}J_{-3}{}^{(2)} \right) g = 0 \tag{8.70}$$

The differential operator $J^{(3)}_m$ appears only at order λ_1^{-n} (n = 1, 2, 3,...). This is similar to the p-spin generalized Kontsevich model without logarithmic term, where the spin 0 equations are described by $J^{(2)}_n$ and the spin non-zero equations by $J^{(3)}_m$ [43].

If we define the differential operators L_m as

$$L_n = \frac{1}{4}J^{(2)}_{2n} \tag{8.71}$$

those L_n have the commutation relations

$$[L_n, L_m] = (n - m)L_{n+m} \tag{8.72}$$

The structure of the generating function for open intersection numbers obeys $W^{(3)}$ algebra [60], which is known to have different operators M_n in addition to above L_n, and these algebraic structures have been studied [5, 119]. It is interesting to consider the open intersection numbers of the moduli space of p-spin curve.

From the duality and replica method in the previous sections, by the expression of $U(\sigma_1, \ldots, \sigma_s)$, these open intersection numbers has been studied [36], and their results agree with the evaluation of (8.56).

Chapter 9
Non-orientable Surfaces from Lie Algebras

9.1 Intersection Numbers from the Lie Algebras $o(N)$ and $sp(N)$

Usually Feyman diagrams related to real symmetric matrices are used for generating non-orientable surfaces. The Euler characteristics of non-orientable surfaces have been derived from real symmetric matrices [68]. There are also studies of integrable systems which satisfies the Drinfeld–Sokolov hierarchy [56], and they are related to non-orientable surfaces [47, 59].

However the lack of HarishChandra formula for such cases, makes it necessary to turn to alternative models. As explained in Sect. 4.3, the HarishChandra formula holds also for the Lie algebras of $o(N)$ and $sp(N)$. The corresponding matrix models $o(N)$ and $sp(N)$ (real antisymmetric matrices and symplectic matrices) produce non-orientable surfaces. So we shall now consider the corresponding matrix models.

$X \in o(2N)$ Lie algebra

When the random matrix X varies over a classical Lie algebra, with Gaussian distribution, the n-point correlation function in an external source may be obtained again in closed form, using the HarishChandra localisation formula [77]. We have discussed in an earlier work such models with an external source [23, 31]. In Sect. 4.3, the duality formula for characteristic polynomials has been discussed.

We rewrite the duality formula of Sect. 4.3,

$$\langle\prod_{\alpha=1}^{k} \det(\lambda_\alpha \cdot \mathrm{I} - X)\rangle_A = \langle\prod_{n=1}^{N} \det(a_n \cdot \mathrm{I} - Y)\rangle_\Lambda \tag{9.1}$$

where X is a $2N \times 2N$ real antisymmetric matrix ($X^t = -X$) and Y is a $2k \times 2k$ real antisymmetric matrix; the eigenvalues of X and Y are thus pure imaginary. The matrix source A is also a $2N \times 2N$ antisymmetric matrix. The matrix Λ is a $2k \times 2k$

E. Brézin and S. Hikami, *Random Matrix Theory with an External Source*,
SpringerBriefs in Mathematical Physics, DOI 10.1007/978-981-10-3316-2_9

antisymmetric matrix, coupled to Y. We assume, without loss of generality, that A and Λ take the canonical form:

By an appropriate tuning of the a_n's, and a corresponding rescaling of Y and Λ, one may generate similarly higher models of type p with the conditions (7.10),

$$Z = \int dY e^{-\frac{1}{p+1}\mathrm{tr}Y^{p+1}+\mathrm{tr}Y\Lambda} \tag{9.2}$$

where p is an odd integer.

The HarishChandra integral for the integral over $g \in SO(2N)$ group, and given real antisymmetric matrices Y and Λ, reads

$$\int_{SO(2N)} dg e^{\mathrm{tr}(gYg^{-1}\Lambda)} = C\frac{\sum_{w\in W}(\det w)\exp\left[2\sum_{j=1}^{N} w(y_j)\lambda_j\right]}{\prod_{1\le j\langle k\le N}(y_j^2-y_k^2)(\lambda_j^2-\lambda_k^2)} \tag{9.3}$$

where $C = (2N-1)!\prod_{j=1}^{2N-1}(2j-1)!$, and w are elements of the Weyl group, which consists here of permutations followed by reflections $(y_i \to \pm y_i\,;\, i = 1, \cdots, N)$ with an even number of sign changes.

For the one point function we obtain from the above formula, when X is a $2N \times 2N$ real antisymmetric random matrix,

$$\begin{aligned}
U(\sigma) &= \frac{1}{2N}\langle \mathrm{tr}e^{\sigma X}\rangle_A \\
&= \frac{1}{2N}\sum_{\alpha=1}^{N}\prod_{\gamma\neq\alpha}^{N}\left(\frac{(a_\alpha+\frac{\sigma}{2})^2-a_\gamma^2}{a_\alpha^2-a_\gamma^2}\right)e^{\sigma a_\alpha+\frac{\sigma^2}{4}} + (\sigma\to-\sigma) \\
&= \frac{1}{N\sigma}\oint\frac{du}{2\pi i}\left(\frac{(u+\frac{\sigma}{2})^2-a_\gamma^2}{u^2-a_\gamma^2}\right)\frac{u}{u+\frac{\sigma}{4}}e^{\sigma u+\frac{\sigma^2}{4}}
\end{aligned} \tag{9.4}$$

where the contour encircles the poles $u = a_\gamma$. Then, shifting $u \to u - \frac{\sigma}{4}$,

$$U(\sigma) = \frac{1}{N\sigma}\oint\frac{du}{2\pi i}\prod_{i=1}^{N}\frac{\left(u-\frac{\sigma}{4}\right)^2-a_i^2}{\left(u+\frac{\sigma}{4}\right)^2-a_i^2}\left(\frac{u-\frac{\sigma}{4}}{u}\right)e^{\sigma u} \tag{9.5}$$

where the contour cycles around all poles at $u = a_i - \frac{\sigma}{4}$ and $u = -a_i - \frac{\sigma}{4}$. Tuning the external source to obtain the p-th degeneracy, one finds

$$U(\sigma) = \frac{1}{N\sigma}\int\frac{du}{2i\pi}e^{-\frac{c}{p+1}\left[(u+\frac{\sigma}{4})^{p+1}-(u-\frac{\sigma}{4})^{p+1}\right]}\left(1-\frac{\sigma}{4u}\right) \tag{9.6}$$

Let us first specialize to specific values of p.

(1) $p = 3$

There are two terms in (9.6); the first term $U(\sigma)^{OR}$ is exactly one-half of $U\left(\frac{\sigma}{2}\right)$ for the GUE (orientable Riemann surfaces). The second term is a new term, and we denote it as the non-orientable part $U(\sigma)^{NO}$, since it is related to non-orientable surfaces with half-integer genus:

$$\begin{aligned} U(\sigma)^{OR} &= \frac{1}{2}U\left(\frac{\sigma}{2}\right) = \frac{1}{12\sqrt{3}}\left[J_{\frac{1}{3}}\left(\frac{1}{12\sqrt{3}}\left(\frac{\sigma}{2}\right)^4\right) + J_{-\frac{1}{3}}\left(\frac{1}{12\sqrt{3}}\left(\frac{\sigma}{2}\right)^4\right)\right] \\ &= \frac{1}{2\cdot 3^{\frac{1}{3}}\left(\frac{\sigma}{2}\right)^{\frac{4}{3}}} Ai\left(-\frac{1}{4\cdot 3^{1/3}}\left(\frac{\sigma}{2}\right)^{\frac{8}{3}}\right) \end{aligned} \tag{9.7}$$

For non-orientable surfaces, from the condition for genus g, characteristics p of spin curves, dimension n, and spin j in (7.22),

$$(p+1)(2g-1) = pn + j + 1 \tag{9.8}$$

we find that the genus g is always a half-integer ($g = \frac{1}{2}, \frac{3}{2}, \frac{5}{2}, \ldots$), and $U(\sigma)^{NO}$ has a series expansion in powers of $\sigma^{n+\frac{1+j}{p}}$. For $p = 3$, we have

$$\begin{aligned} U(\sigma)^{NO} &= \frac{1}{4}\oint \frac{du}{2\pi i}\frac{1}{u}e^{-\frac{\sigma}{2}u^3 - \frac{\sigma^3}{32}u} \\ &= \frac{1}{12}\oint \frac{dx}{2\pi i}\frac{1}{x}e^{-x-\frac{2^{1/3}}{32}\sigma^{8/3}x^{1/3}} \\ &= \mathrm{Re}\left\{\frac{1}{12i\pi}\int_0^\infty \frac{1}{x}e^{-x-\frac{1}{4}e^{\frac{2\pi i}{3}}\left(\frac{\sigma}{2}\right)^{8/3}x^{1/3}}\right\} \end{aligned} \tag{9.9}$$

This function is expanded as

$$\begin{aligned} U(\sigma)^{NO} &= \mathrm{Re}\left\{\frac{1}{12i\pi}\int_0^\infty dx\frac{1}{x}e^{-x}\sum_{n=0}^{\infty}\frac{1}{n!}\left(-\frac{1}{4}e^{\frac{2\pi i}{3}}\left(\frac{\sigma}{2}\right)^{8/3}x^{1/3}\right)^n\right\} \\ &= -\frac{1}{\pi}\frac{1}{48}\left(\frac{\sigma}{2}\right)^{\frac{8}{3}}\left(\sin\frac{2\pi}{3}\right)\Gamma\left(\frac{1}{3}\right) + \frac{1}{\pi}\frac{1}{384}\left(\frac{\sigma}{2}\right)^{\frac{16}{3}}\left(\sin\frac{4\pi}{3}\right)\Gamma\left(\frac{2}{3}\right) \\ &\quad - \frac{1}{\pi}\frac{1}{3!\cdot 12\cdot 4^3}\left(\frac{\sigma}{2}\right)^8(\sin 2\pi) + \frac{1}{\pi}\frac{1}{4!\cdot 12\cdot 4^4}\left(\frac{\sigma}{2}\right)^{\frac{32}{3}}\left(\sin\frac{8\pi}{3}\right)\Gamma\left(\frac{4}{3}\right) - \cdots \end{aligned} \tag{9.10}$$

Using Airy functions, the $p = 3$ case is expressed as

$$U(\sigma) = \frac{1}{2\cdot 3^{1/3}\left(\frac{\sigma}{2}\right)^{4/3}}Ai(x) - \frac{1}{4}\int_0^x dx' Ai(x') \tag{9.11}$$

with $x = -\frac{1}{4 \cdot 3^{1/3}} \left(\frac{\sigma}{2}\right)^{8/3}$. The Airy function $Ai(z)$ and the integral of the Airy function may be expanded as

$$Ai(z) = \frac{\pi}{3^{2/3}} \sum_{n=0}^{\infty} \frac{1}{n!\Gamma\left(n+\frac{2}{3}\right)} \left(\frac{1}{3}\right)^{2n} z^{3n} - \frac{\pi}{3^{4/3}} \sum_{0}^{\infty} \frac{1}{n!\Gamma\left(n+\frac{4}{3}\right)} \left(\frac{1}{3}\right)^{2n} z^{3n+1} \tag{9.12}$$

$$\int_0^z Ai(t)dt = \frac{\pi}{3^{2/3}\Gamma\left(\frac{2}{3}\right)} z - \frac{\pi}{3^{4/3} \cdot 2\Gamma\left(\frac{4}{3}\right)} z^2 + \frac{\pi}{36 \cdot 3^{2/3}\Gamma\left(\frac{5}{3}\right)} z^4 + \cdots \tag{9.13}$$

Inserting these expansions, we have for $p = 3$, $\frac{\pi}{\sin(\frac{\pi}{3})} = \Gamma\left(\frac{1}{3}\right)\Gamma\left(\frac{2}{3}\right)$.

$$\begin{aligned} U(\sigma) &= \frac{\pi}{24\Gamma\left(\frac{1}{3}\right)} \left(\frac{\sigma}{2}\right)^{4/3} - \frac{\pi}{108 \cdot 64\Gamma\left(\frac{2}{3}\right)} \left(\frac{\sigma}{2}\right)^{20/3} + \cdots \\ &+ \left[\frac{\pi}{48\Gamma\left(\frac{2}{3}\right)} \left(\frac{\sigma}{2}\right)^{8/3} + \frac{\pi}{384\Gamma\left(\frac{1}{3}\right)} \left(\frac{\sigma}{2}\right)^{16/3} - \frac{\pi}{864 \cdot 4^4\Gamma(\frac{2}{3})} \left(\frac{\sigma}{2}\right)^{32/3} + \cdots\right] \end{aligned} \tag{9.14}$$

$$\begin{aligned} U(\sigma) &= \langle\tau_{1,0}\rangle_{g=1}\Gamma\left(1-\frac{1}{3}\right)\left(\frac{\sigma}{2}\right)^{1+\frac{1}{3}} + \langle\tau_{6,1}\rangle_{g=3}\Gamma\left(1-\frac{2}{3}\right)3^2\left(\frac{\sigma}{2}\right)^{6+\frac{2}{3}} + \cdots \\ &+ \left[\langle\tau_{2,1}\rangle_{g=3/2}\Gamma\left(1-\frac{2}{3}\right)3^2\left(\frac{\sigma}{2}\right)^{2+\frac{2}{3}} + \langle\tau_{5,0}\rangle_{g=5/2}\Gamma\left(1-\frac{1}{3}\right)3^4\left(\frac{\sigma}{2}\right)^{16/3}\right. \\ &\left. + \langle\tau_{10,1}\rangle_{g=9/2}\Gamma\left(1-\frac{2}{3}\right)3^8\left(\frac{\sigma}{2}\right)^{32/3} + \cdots\right] \end{aligned} \tag{9.15}$$

We have for $p = 3$,

$$U(\sigma)^{NO} = \frac{1}{12}y^2\Gamma\left(1-\frac{2}{3}\right) + \frac{1}{24}y^4\Gamma\left(1-\frac{1}{3}\right) + \frac{1}{864}y^8\Gamma\left(1-\frac{2}{3}\right) + \cdots \tag{9.16}$$

We have obtained for $p = 3$ the explicit intersection numbers for non-orientable surfaces with one marked point. The intersection number $\langle\tau_{2,1}\rangle_{g=3/2}$ corresponds to a cross-capped torus. For $g = 1/2$ we are dealing with the topology of the projective plane but for this case, the intersection numbers $\langle\tau_{0,1}^2\rangle_{g=1/2}$ are present only beyond the two marked points level [31]. We have

$$\langle\tau_{1,0}\rangle_{g=1} = \frac{1}{24}, \quad \langle\tau_{2,1}\rangle_{g=\frac{3}{2}} = \frac{1}{864}, \cdots \tag{9.17}$$

(2) general p

Using the binomial expansion, one finds ($y = \frac{1}{2}\left(\frac{\sigma}{2}\right)^{1+\frac{1}{p}}$)

$$U(\sigma) = -\frac{1}{4ypN}\int dt t^{\frac{1}{p}-1} e^{-t}\left[1 - \frac{p(p-1)}{6} y^2 t^{1-\frac{2}{p}} + \cdots\right] \times \left[1 + yt^{-\frac{1}{p}}\right] \tag{9.18}$$

This is again the sum of two contributions, orientable (OR) and non-orientable (NO). The odd powers in y correspond to the orientable contribution, which is the same as for the unitary case; the even powers in y correspond to the non-orientable case:

$$U(\sigma) = U(\sigma)^{OR} + U(\sigma)^{NO} \tag{9.19}$$

$U(\sigma)^{OR}$ is same as GUE with σ replaced by $\sigma/2$.

The first term in the above series expansion is divergent, and it should be regularized. Discarding this divergent term, we give the series expansion up to order y^8 (neglecting the phase factor $\sin\left(\frac{2\pi m}{p}\right)$),

$$\begin{aligned}
U(\sigma)^{NO} &= \frac{y^2}{24}(p-1)\Gamma\left(1-\frac{2}{p}\right) + \frac{y^4}{6!}(p-1)(p^2-5p+1)\Gamma\left(1-\frac{4}{p}\right) \\
&+ \frac{y^6}{7!\cdot 9}(p-1)(p-3)(4p^3-23p^2-2p-6)\Gamma\left(1-\frac{6}{p}\right) \\
&+ \frac{y^8}{7!3^3\cdot 10}(p-1)(9p^6-121p^5+435p^4-317p^3 \\
&-167p^2-471p-43)\Gamma\left(1-\frac{8}{p}\right) + O(y^{10})
\end{aligned} \tag{9.20}$$

From this genus expansion, one obtains the intersection numbers of p-spin curves for non-orientable surfaces.

(3) $p = -1$

We now consider the limit, $p \to -1$, which we found to be related to the virtual Euler characteristics. When we set $p = -1$ in (9.20), the Γ function term becomes an integer, and this agrees with the intersection number of $\langle\tau_{1,0}\rangle_g$, which gives a factor $\Gamma\left(1-\frac{1}{p}\right) = \Gamma(2) = 1$ for the spin zero. We obtain

$$U(\sigma)^{NO} = -\frac{1}{24}(2y)^2 - \frac{7}{240}(2y)^4 - \frac{31}{504}(2y)^6 - \frac{127}{480}(2y)^8 + \cdots \tag{9.21}$$

This series agrees precisely with the series expansion

$$U(\sigma)^{NO} = -\sum_{\hat{g}=1}^{\infty}\frac{1}{2\hat{g}}\left(2^{2\hat{g}-2} - \frac{1}{2}\right) B_{\hat{g}}(2y)^{2\hat{g}} \tag{9.22}$$

where $B_{\hat{g}}$ is a Bernoulli number in (9.30). The coefficient of $(2y)^{2\hat{g}}$ is the same as for the virtual Euler characteristics of the moduli space of real algebraic curves for genus g and one marked point, which was derived from the Penner model of the real symmetric matrix by Goulden et al. [68]. (We use for the half genuses in the list, $\frac{1}{2}$, 1, $\frac{3}{2}$, 2, ...for a projective plane, Klein bottle, cross-capped torus, doubly cross-capped torus, ..., with the notation $\hat{g} = 1$, $\hat{g} = 2$, $\hat{g} = 3$,$\hat{g} = 4$, ..., respectively [82], and this is the reason for the appearance of the $(2y)^{2\hat{g}}$ factor in (9.21)).

Since we derived this from the antisymmetric $o(2N)$ Lie algebra, the coincidence between $o(2N)$ lie algebra and GOE for the virtual Euler characteristics seems remarkable.

$$\chi^{NO}(\bar{M}_{g,1}) = \frac{1}{2g}\left(\frac{1}{2} - 2^{2g-2}\right) B_g. \tag{9.23}$$

This result may be obtained analytically to all orders. We now derive this result from the integral form (2.23) replacing c by N. With $p = -1$, it becomes

$$U(\sigma) = -\frac{1}{4N\sigma}\int du \left(\frac{u-\sigma}{u+\sigma}\right)^N \left(1 + \frac{\sigma}{u}\right) \tag{9.24}$$

With the change of variable $u \to \sigma u$,

$$U(s) = -\frac{1}{4N}\int du \left(\frac{u-1}{u+1}\right)^N \left(1 + \frac{1}{u}\right) \tag{9.25}$$

We divide it into two parts, $U(\sigma)^{OR}$ and $U(\sigma)^{NO}$,

$$U(\sigma)^{OR} = -\frac{1}{4N}\int du \left(\frac{u-1}{u+1}\right)^N \tag{9.26}$$

$$U(\sigma)^{NO} = -\frac{1}{4N}\int du \left(\frac{u-1}{u+1}\right)^N \frac{1}{u} \tag{9.27}$$

We use the same change of variables as for the unitary case in (7.108),

$$U(\sigma)^{OR} = \frac{1}{2N}\int dy e^{-Ny}\frac{e^{-y}}{(1-e^{-y})^2} \tag{9.28}$$

$$U(\sigma)^{NO} = \frac{1}{2N}\int dy e^{-Ny}\frac{e^{-y}}{(1-e^{-y})^2}\left(\frac{1-e^{-y}}{1+e^{-y}}\right) = \frac{1}{4N}\int dy e^{-Ny}\left[\frac{1}{1-e^{-y}} - \frac{1}{1+e^{-y}}\right] \tag{9.29}$$

It is interesting to note that both the Bose-Einstein and the Fermi-Dirac distributions enter in the above integrand (9.29).

If we use the expansions,

$$\frac{1}{1-e^{-y}} = \frac{1}{y} + \frac{1}{2} + \sum_{n=1}^{\infty}(-1)^{n-1}\frac{B_n}{(2n)!}y^{2n-1}$$
$$\frac{1}{1+e^{-y}} = \frac{1}{2} + \sum_{n=1}^{\infty}\frac{(-1)^{n-1}(2^{2n}-1)}{(2n)!}B_n y^{2n-1} \tag{9.30}$$

then they become

$$U(\sigma)^{OR} = \frac{1}{2N}\int dy\frac{1}{y^2}e^{-Ny} - \frac{1}{2}\sum_{n=1}^{\infty}(-1)^{n-1}\frac{B_n}{2n}\frac{1}{N^{2n}}$$
$$U(\sigma)^{NO} = \frac{1}{4N}\int dye^{-Ny} + \frac{1}{4}\sum_{n=1}^{\infty}(-1)^{n-1}\frac{B_n}{2n}\frac{1}{N^{2n+1}}$$
$$-\frac{1}{4}\sum_{n=1}^{\infty}(-1)^{n-1}\frac{(2^{2n}-1)}{2n}B_n\frac{1}{N^{2n+1}}$$
$$= \frac{1}{4N}\int dy\frac{e^{-Ny}}{y} + \frac{1}{4}\sum_{n=1}^{\infty}(-1)^{n-1}\frac{(2-2^{2n})B_n}{2n}\frac{1}{N^{2n+1}} \tag{9.31}$$

We now get from the above expression (replacing n by g),

$$\chi^{OR}(\bar{M}_{g,1}) = -\frac{1}{2}\zeta(1-2g) = -\frac{1}{2}\frac{(-1)^g B_g}{2g},$$
$$\chi^{NO}(\bar{M}_{g,1}) = (-1)^{g-1}\frac{1}{2g}(2^{2g-2}-2^{-1})B_g \tag{9.32}$$

For s marked points, the result obtained from the real symmetric matrix Penner model by Goulden et. al. [68] is

$$\chi^{NO}(\bar{M}_{g,s}) = (-1)^s\frac{1}{2}\frac{(2g+s-2)!(2^{2g-1}-1)}{(2g)!s!}B_g \tag{9.33}$$

This result may be obtained by applying the Eq. (9.32) [35]. In this o(2N) model, we have the following condition, the same as for Riemann surfaces with spin j and s-marked points

$$(p+1)(2g-2+s) = p\sum_{i=1}^{s}n_i + \sum_{i=1}^{s}j_i + s \tag{9.34}$$

However, we have to consider half integer values of the genus g to represent non-orientable surfaces [31].

$X \in o(2N+1)$ Lie algebra

For $so(2N+1)$ Lie algebra, the matrix X is $X = h_1 v \oplus h_2 v \oplus \cdots h_N v \oplus 0$. The measure is $V(H)^2$, $V(H) = \prod_{1\le j\le N}(h_j^2 - h_k^2)\prod_{j=1}^N h_j$. The HarishChandra formula is

$$I = \int_{SO(2N+1)} e^{\mathrm{tr}(gag^{-1}b)} dg = C_{G(N)} \frac{\sum\limits_{w\in G(N)} (\det w)\exp\left(2\sum\limits_{j=1}^N w(a_j)b_j\right)}{\prod\limits_{1\le j\le k\le N}(a_j^2 - a_k^2)(b_j^2 - b_k^2)\prod\limits_{j=1}^N a_j b_j} \tag{9.35}$$

with $C_{G(n)} = \prod_{j=1}^N (2j-1)!\prod_{j=2N}^{4N-1} j!$. Compared with the $o(2N)$ case, this formula differs from (9.3) by the presence of the term $\prod a_j b_j$ in the denominator. For the one point function, we have

$$U(\sigma) = \frac{1}{N}\sum_{\alpha=1}^N \int_{-\infty}^{\infty}\prod_{i=1}^N d\lambda_i \frac{\prod(\lambda_i^2 - \lambda_j^2)\prod\lambda_k}{\prod(a_i^2 - a_j^2)\prod a_k} e^{-\sum\lambda_i^2 + \sigma\lambda_\alpha + 2\sum a_i\lambda_i} \tag{9.36}$$

This sum of integrals may be written as a contour integral, which collects poles at $u = a_i^2$,

$$\begin{aligned} U(\sigma) &= \oint_{\{u=a_i^2\}} \frac{du}{2\pi i}\prod_{j=1}^N \frac{(\sqrt{u}+\sigma)^2 - a_j^2}{u - a_j^2}\frac{1}{(\sqrt{u}+\sigma)^2 - u}\left(1+\frac{\sigma}{\sqrt{u}}\right)e^{\sigma^2 + 2\sigma\sqrt{u}} \\ &= \frac{2}{\sigma}\oint \frac{dv}{2\pi i}\prod_{j=1}^N \frac{(v+\sigma)^2 - a_j^2}{v^2 - a_j^2}\frac{v+\sigma}{\sigma+2v}e^{\sigma^2+2\sigma v} \\ &= \frac{1}{\sigma}\oint \frac{dv}{2\pi i}\prod_{j=1}^N \frac{\left(v+\frac{\sigma}{2}\right)^2 - a_j^2}{\left(v-\frac{\sigma}{2}\right)^2 - a_j^2}\left(1+\frac{\sigma}{2v}\right)e^{\sigma v} \end{aligned} \tag{9.37}$$

By tuning to the p-th degeneracy, we obtain

$$U(\sigma) = \frac{1}{\sigma}\oint \frac{du}{2\pi i} e^{-\frac{1}{p+1}\left(\left(u+\frac{\sigma}{2}\right)^{p+1} - \left(u-\frac{\sigma}{2}\right)^{p+1}\right)}\left(1+\frac{\sigma}{2u}\right) \tag{9.38}$$

This takes the same form as for the $o(2N)$ case.

$X \in sp(N)$ Lie algebra

The Haar measure of $sp(N)$ is $\Delta(\lambda)^2$, with $\Delta(\lambda) = \prod_{i<j}(\lambda_i^2 - \lambda_j^2)\prod_k \lambda_k$. The HarishChandra formula for $sp(N)$ reads [23]

$$\begin{aligned} I &= \int_G e^{\langle Ad(g)\cdot a|b\rangle} dg = \frac{\sum_{w\in W}(\det w)e^{\langle w\cdot a|b\rangle}}{\Delta(a)\Delta(b)} \\ &= C\frac{\det[2\sinh(2a_ib_j)]}{\prod(a_i^2-a_j^2)(b_i^2-b_j^2)\prod(a_kb_k)} \end{aligned} \tag{9.39}$$

For the one point function, we have

$$\begin{aligned} U(\sigma) &= \frac{1}{N}\sum_{\alpha=1}^{N}\int_{-\infty}^{\infty}\prod_{i=1}^{N}d\lambda_i \frac{\prod\limits_{1\le i\langle j\le N}(\lambda_i^2-\lambda_j^2)\prod\limits_{1\le k\le N}\lambda_k}{\prod\limits_{1\le i\langle j\le N}(a_i^2-a_j^2)\prod\limits_{1\le k\le N}a_k} e^{-\sum\lambda_i^2+\sigma\lambda_\alpha+2\sum a_i\lambda_i} \\ &= \oint\frac{du}{2\pi i}\prod_{j=1}^{N}\frac{(\sqrt{u}+\sigma)^2-a_j^2}{u-a_j^2}\frac{1}{(\sqrt{u}+\sigma)^2-u}\left(1+\frac{\sigma}{\sqrt{u}}\right)e^{\sigma^2+2\sigma\sqrt{u}} \\ &= \frac{2}{\sigma}\oint\frac{dv}{2\pi i}\prod_{j=1}^{N}\frac{(v+\sigma)^2-a_j^2}{v^2-a_j^2}\frac{v+\sigma}{\sigma+2v}e^{\sigma^2+2\sigma v} \\ &= \frac{1}{\sigma}\oint\frac{dv}{2\pi i}\prod_{j=1}^{N}\frac{\left(v+\frac{\sigma}{2}\right)^2-a_j^2}{\left(v-\frac{\sigma}{2}\right)^2-a_j^2}\left(1+\frac{\sigma}{2v}\right)e^{\sigma v} \end{aligned} \tag{9.40}$$

where we have shifted $v\to v-\frac{\sigma}{2}$ and $a_\gamma\to a_\gamma/2$. This expression is identical to that of the $o(2N)$ case, when we change $v\to 2v$, up to a factor 2. Note that we do not need to consider the replacement $\sigma\to\frac{\sigma}{2}$ as in the $o(2N)$ case. The first term of the expression is the same as for GUE. By tuning the a_γ to the p-th case, we have

$$U(\sigma) = \frac{1}{\sigma}\oint\frac{du}{2\pi i}e^{-\frac{1}{p+1}\left(\left(u+\frac{\sigma}{2}\right)^{p+1}-\left(u-\frac{\sigma}{2}\right)^{p+1}\right)}\left(1+\frac{\sigma}{2u}\right) \tag{9.41}$$

We write these two terms as $U(\sigma)=U(\sigma)^{OR}+U(\sigma)^{NO}$. It is then obvious that we obtain the same intersection numbers and virtual Euler characteristics as with the $o(2N)$ case.

Chapter 10
Gromov–Witten Invariants, $\mathbf{P}^1$ Model

The intersection numbers of p-spin curves is a simple example of more general Gromov–Witten invariants, where the manifold X is a point.

Gromov–Witten invariants are the subjects of active research. Here one considers only the simplest example of $\mathbf{P}^1$ case, which is related to this present book [36]. $\mathbf{P}^1$ means the complex projection manifold $CP^1 = SU(2)/U(1)$, and represents a sphere of topology.

The definition of the Gromov–Witten invariants of the moduli space $\overline{\mathscr{M}}_{g,s}(\mathbf{P}^1, d)$, for genus g, s marked points, manifold $X = \mathbf{P}^1$ and degree d, is given by

$$\left\langle \prod_{i=1}^{l} \tau_{n_i}(1) \prod_{i=l+1}^{s} \tau_{n_i}(\omega) \right\rangle_{g,d} = \int_{\overline{\mathscr{M}}_{g,s}(P^1,d)} \prod_{i=1}^{l} c_i(\mathscr{L}_i)^{n_i} \prod_{i=l+1}^{s} c_i(\mathscr{L}_i)^{n_i} ev_i(\omega) \tag{10.1}$$

where $\omega \in H^2(P^1, Q)$ and $1 \in H^0(P^1, Q)$, and $ev_i^*(\omega) \in H^2(\overline{\mathscr{M}}_{g,n}(X, d), Q)$, ev_i is morphism: $\overline{\mathscr{M}}_{g,n}(X) \to X$. H^2 and H^0 are cohomological classes. When X is P^1, it vanishes unless the condition for the moduli space $\overline{\mathscr{M}}_{g,s}(X, d)$ is satisfied,

$$2(g-1) + s = \sum_{i=1}^{s} n_i + \sum_{i=1}^{s} q_i - 2d, \tag{10.2}$$

where d is degree, n_i is parameter defined as above, and q_i (i = 1, ...,s) is $U(1)$ charge of instantons. In P^1, there are cohomological two classes $(1, \omega)$. The ω class, $\omega \in H^2(X, Q)$, is related to the instanton, $\pi_2(X) = Z$ (integers). We consider here only stationary sector of the Gromov–Witten theory, i.e. ω class, which is integral involving only the descendants of ω, since it is fundamental.

The matrix model for $\mathbf{P}^1$ has been studied with a logarithmic potential through Virasoro equations [57]. By the duality (mirror symmetry), different matrix model can be considered, which is

E. Brézin and S. Hikami, *Random Matrix Theory with an External Source*,
SpringerBriefs in Mathematical Physics, DOI 10.1007/978-981-10-3316-2_10

$$H_{GL} = \mathrm{tr}\left(M + \frac{1}{M}\right) \tag{10.3}$$

This is due to the form of the Ginzburg–Landau Hamiltonian H_{GL} (a quantum ring for the genus zero) for $\mathbf{P}^n$ is

$$H_{GL} = M_1 + \cdots + M_n + \prod_{i=1}^{n} \frac{1}{M_i} \tag{10.4}$$

Since $X = \mathbf{P}^1$ is a sphere, the periodic covers are important, and therefore instead of M, it is useful to consider

$$H = \mathrm{tr}(e^M + e^{-M}) + \mathrm{trMA} \tag{10.5}$$

This form resembles to the sine-Gordon model. The $\mathbf{P}^1$ model (or supersymmetric $\mathbf{P}^1$ model) is well known as a toy model of four dimensional gauge fields. It is asymptotically free, namely the β-function of the renormalization group is negative at the origin, and it has an instanton.

The intersection numbers of Gromov–Witten theory are holomorphic maps from Riemann surface onto a target manifold, and p-spin curves corresponds to the point target space. It corresponds to the supersymmetric nonlinear sigma model in two dimensions, with a target manifold $\mathbf{P}^1$, which is known to be asymptotic free. We will apply duality and replica method in previous sections to Gromov–Witten theory of $\mathbf{P}^1$.

Recently, this Gromov–Witten theory for $\mathbf{P}^1$ has been studied from various point of views [57]. The intersection numbers of stationary sector, ω class, for $\mathbf{P}^1$ have been evaluated in [104, 109], and we will compare results with them.

The definition of the intersection numbers for s-marked point for stationary sector is [108]

$$\left\langle \prod_{i=1}^{s} \tau_{n_i}(\omega) \right\rangle_{g,d} = \int_{\overline{\mathscr{M}}_{g,s}} \prod_{i=1}^{s} c_1(\mathscr{L}_i)^{n_i} ev_i(\omega) \tag{10.6}$$

which may be compared the expression for p spin curve with the top Chern class $c_D(\mathscr{V})$ in (7.3),

$$\left\langle \prod_{i=1}^{s} \tau_{n_i,m_i} \right\rangle_{g} = \frac{1}{p^g} \int_{\overline{\mathscr{M}}^{1/p}_{g,s}} \prod_{i=1}^{s} c_1(\mathscr{L}_i)^{n_i} c_D(\mathscr{V}) \tag{10.7}$$

The matrix model for the Gromov–Witten theory of $\mathbf{P}^1$ is similar to the Kontsevich matrix model [2], which we will investigate below by applying the previous method based on external source and duality. The partition function is

$$Z_{\mathbf{P}^1}(\Lambda) = \int dB e^{-\frac{1}{g_s}\mathrm{tr}(e^B + qe^{-B} - B\Lambda)} \tag{10.8}$$

where B is an $n \times n$ Hermitian matrix, and g_s and q are coupling constants. The partition function $Z_{\mathbf{P}^1}(\Lambda)$ is characterized by the parameters t_n

$$t_n = \frac{1}{n}\mathrm{tr}\frac{1}{\Lambda^n} \tag{10.9}$$

which correspond to Toda times. The model is similar to the Kontsevich model, but the fractional power seen in p-th Airy matrix model is not necessary. This model has the form of a superpotential $V(x) = e^x + qe^{-x}$ in genus zero, where $q = e^{t_{0,Q}}$, which appears in $\mathbf{P}^1$ problem [57]. This model is then similar to the sine-Gordon model, and it seems reasonable since the supersymmetric $\mathbf{P}^1$ model has a close connection to the sine-Gordon model for generating a mass.

Let consider the simplest case by taking $n = 1$, (n is the size of the matrix B, and B is x), in (10.8); we find

$$Z_{P^1}(\lambda) = \int_{-\infty}^{\infty} dx e^{-\frac{1}{g_s}(e^x + qe^{-x} - x\lambda)} \tag{10.10}$$

When g_s is small ($g_s \sim \frac{1}{N}, N \to \infty$), we use a saddle point approximation and expand around the saddle point $x_c = \log[(\lambda \pm \sqrt{\lambda^2 + 4q})/2]$:

$$g_s F = -\lambda \log\frac{\lambda \pm \sqrt{\lambda^2 + 4q}}{2} \pm \sqrt{\lambda^2 + 4q} \tag{10.11}$$

Taking $x_c = \log[(\lambda + \sqrt{\lambda^2 + 4q})/2]$, we find the expansion,

$$\begin{aligned} g_s F_+ &= -\lambda \log\lambda + \lambda + \lambda \sum_{m=1}^{\infty}\left(\frac{\Gamma(\frac{3}{2})}{m!\Gamma(\frac{3}{2} - m)} - \frac{(2m-1)!!}{(2m)!!(2m)}\right)\left(\frac{4q}{\lambda^2}\right)^m \\ &= -\lambda \log\lambda + \lambda + \frac{q}{\lambda} - \frac{7}{2}\frac{q^2}{\lambda^3} + \frac{2}{3}\frac{q^3}{\lambda^5} + O(q^5) \end{aligned} \tag{10.12}$$

The Gromov–Witten theory for $\mathbf{P}^1$ corresponds to the supersymmetric nonlinear sigma model in two dimensions for $\mathbf{P}^1$ [41]. The form of F is consistent with a supersymmetric $\mathbf{P}^1$ model with a curve of marginal stability, by identifying λ as a mass m and q as a normalized scaling momentum $\tilde{\Lambda}$.

Since the form of the matrix model in (10.8) resembles Kontsevich model or the p-th higher Airy matrix model with an external source, we can apply the duality formula of Gaussian matrix model with an external source tuned to provide the potential (10.8). Using the same replica method as in the p-th higher Airy matrix model, we start with $U(\sigma)$. Using the duality formula of external source a_j, we obtain the general polynomial of B in (7.126),

$$
\begin{aligned}
\left\langle \prod_{i=1}^{N} \det(a_j - iB) \right\rangle_{\Lambda} &= \int dB e^{-\frac{1}{2g_s}\mathrm{tr}B^2 + \frac{1}{g_s}\mathrm{tr}B\Lambda + \sum_{i=1}^{N} \mathrm{tr}\log(a_i - iB)} \\
&= \int dB e^{-\frac{1}{2g_s}\mathrm{tr}B^2 + \frac{1}{g_s}\mathrm{tr}B\Lambda + \sum_j \log a_j + \sum_j \mathrm{tr}\log\left(1 - \frac{iB}{a_j}\right)} \\
&= \int dB e^{\frac{1}{g_s}[\mathrm{tr}B\Lambda - \mathrm{tr}(e^B + qe^{-B})]}
\end{aligned} \tag{10.13}
$$

where we put $c_{m_1} = (1 + q(-1)^{m_1})/m_1!$ in (7.126), by the tuning of the external source a_j. We have

$$
U(\sigma) = \frac{e^{\frac{1}{2}\sigma^2}}{\sigma} \oint \frac{du}{2\pi i} \prod_{j=1}^{N} \frac{a_j - (u + \sigma)}{a_j - u} e^{u\sigma} \tag{10.14}
$$

Taking the same conditions for a_j as in (10.13), we find

$$
\begin{aligned}
U(\sigma) &= \frac{1}{\sigma} \oint \frac{du}{2\pi i} e^{-\frac{1}{g_s}[(e^{u+\sigma} + qe^{-(u+\sigma)}) - (e^u + qe^{-u})]} \\
&= \frac{1}{\sigma} \oint \frac{du}{2\pi i} e^{-\frac{2}{g_s}(\sinh\frac{\sigma}{2})(e^{u+\frac{\sigma}{2}} - qe^{-u-\frac{\sigma}{2}})}
\end{aligned} \tag{10.15}
$$

By the change of variable, $x = e^u$, we have

$$
U(\sigma) = \frac{1}{\sigma} \oint \frac{dx}{2\pi i} \frac{1}{x} e^{-\frac{2}{g_s}\sinh\frac{\sigma}{2}(xe^{\frac{\sigma}{2}} - qx^{-1}e^{-\frac{\sigma}{2}})} \tag{10.16}
$$

This integral is done by taking the residue at $x = 0$,

$$
\begin{aligned}
U(\sigma) &= \frac{1}{\sigma} \sum_{d=0}^{\infty} \frac{q^d}{d!d!} (-1)^d \left(\frac{2}{g_s} \sinh\frac{\sigma}{2} \right)^{2d} \\
&= \frac{1}{\sigma} J_0 \left(\frac{2\sqrt{q}}{g_s} \sinh\frac{\sigma}{2} \right)
\end{aligned} \tag{10.17}
$$

where J_0 is Bessel function of order zero. The constant g_s is of order $\frac{1}{N}$, and it is scaled by the change $\sigma \to g_s\sigma$, which gives a $\frac{1}{N}$ genus expansion. Therefore, we are able to put $g_s = 1$, and the genus g is extracted from the power of σ as $\sigma^{2g+2d-1}$.

$$
U(\sigma) = \frac{1}{\sigma}\left[1 - q\left(\sigma^2 + \frac{1}{12}\sigma^4 + \cdots\right) + q^2\left(\frac{1}{16}\sigma^4 + \cdots\right) + O(q^3)\right] \tag{10.18}
$$

The genus zero (g = 0) is

$$U(\sigma)_{g=0} = \sum_{d=0}^{\infty} \frac{(-1)^d}{d!d!} q^d \sigma^{2d-1} \tag{10.19}$$

The Gromov–Witten theory is a generalization of the intersection theory of moduli space of curves. The moduli space (stack) of stable maps is denoted by $M_{g,s}(X, d)$, where g is genus, s is marked point, and d is degree. The condition of this moduli space is for $X = \mathbf{P}^1$,

$$\dim M_{g,s}(\mathbf{P}^1, d) = 2d + 2(g-1) + s \tag{10.20}$$

where $\dim M_{g,s} = \sum_{i=1}^{s}(n_i + q_i)$. The instanton of $\mathbf{P}^1$ has to be counted, and q_i is $U(1)$ charge of the field. Thus we have

$$2(g-1) + s = \sum_{i=1}^{s} n_i + \sum_{i=1}^{s} q_i - 2d \tag{10.21}$$

Comparing this with the condition for p-spin curves with s marked point in (7.1), in which

$$2(g-1) + s = \sum_{j=1}^{s} n_j + \frac{1}{p}\sum_{i=1}^{s} j_i - \frac{2(g-1)}{p} \tag{10.22}$$

we find the U(1) charge q_i (i is index of a marked point), corresponds to the spin j_i, ($j_i = 0, \ldots, p-1$). The degree d corresponds to $\frac{(g-1)}{p}$ for the case of p-spin curves. The number 2 in front of d appears according to $c_1(\mathbf{P}^1) = 2$. For one marked point, $s = 1$, we have

$$n + q_1 = 2g + 2d - 1 \tag{10.23}$$

and the Free energy F has a following expansion by above condition,

$$q_s F = \sum_{n,q_1}\sum_{d} C_{n,q_1}(d) \frac{q^d}{\lambda^{n+q_1}} \tag{10.24}$$

and we find that the expansion in (10.12) agrees with above formula for $g = 0$ and the expansion of (10.17) for $g \geq 0$ agrees the form of (10.24). From (10.23), we obtain the expression of Gromov Witten invariants for genus zero with $q_1 = 1$ as

$$\langle \tau_{2n-2} \rangle_{g=0} = \frac{1}{(n!)^2} \tag{10.25}$$

For higher genus, we obtain terms of all genus g of one marked point from (10.18). The first few terms are

$$\langle \tau_{2n}(\omega)\rangle_{g=1} = \frac{1}{24(n!)^2}(2n-1)$$
$$\langle \tau_{2n}(\omega)\rangle_{g=2} = \frac{1}{2^7 3^2 5 (n!)^2} n^2(2n-3)(10n-17)$$
$$\langle \tau_{2n}(\omega)\rangle_{g=3} = \frac{1}{2^{10}3^4 5\cdot 7}\frac{1}{(n!)^2} n^2(n-1)^2(2n-5)(140n^2-784n+1101) \tag{10.26}$$

These results are obtained from (10.23) by collecting the coefficient of σ^{2d}, and making the shift $2d \to 2d-2g+1$ in the expansion of $(2\sinh\frac{\sigma}{2})^{2d}$. The obtained Gromov–Witten invariant for stationary sector is consistent with the result of [109], which gives

$$\sum_{g=0}^{\infty} \sigma^{2g}\langle \tau_{2g-2+2d}(\omega)\rangle = \frac{1}{(d!)^2}\left(\frac{\sinh\frac{\sigma}{2}}{\frac{\sigma}{2}}\right)^{2d-1}. \tag{10.27}$$

Since we have used a replica method, $n \to 0$ (n is the size of the matrix Λ), the form of F with one marked point, for the matrix model of (10.8) is considered as

$$q_s F = \sum_{n,q_1}\sum_{d} C_{n,q_1}(d) q^d \mathrm{tr}\left(\frac{1}{\Lambda^{n+q_1}}\right) \tag{10.28}$$

with the condition (10.23).

The correlation function for s-marked points, $U(\sigma_1,\ldots,\sigma_n)$ is

$$U(\sigma_1,\ldots,\sigma_s) = \oint \prod_{k=1}^{s}\frac{du_k}{2\pi i}\prod \det\left(\frac{1}{u_i-u_j+\sigma_i}\right) e^{-\sum_{i=1}^{s}[(e^{u_i+\sigma_i}+qe^{-(u_i+\sigma_i)})-(e^{u_i}+qe^{-u_i})]} \tag{10.29}$$

The connected part is realized as a maximum cyclic loop in the determinant. For two point function, with $z_i = e^{u_i}$, we have

$$U(\sigma_1,\sigma_2) = \oint \frac{dz_1 dz_2}{(2\pi i)^2}\frac{1}{z_1 z_2}\frac{1}{(z_1-z_2+\sigma_1)(z_2-z_1+\sigma_2)} \times e^{-(\sinh\frac{\sigma_1}{2})(z_1-qz_1^{-1})-(\sinh\frac{\sigma_2}{2})(z_2-qz_2^{-1})} \tag{10.30}$$

The shifts $z_1 \to \sigma_1 z_1$ and $z_2 \to \sigma_2 z_2$ give

$$\frac{1}{\sigma_1(z_1+1)-\sigma_2 z_2}\cdot\frac{1}{\sigma_2(z_2+1)-\sigma_1 z_1} = -\frac{1}{\sigma_2^2 z_2(z_2+1)}\sum_{m=0}^{\infty}\left(\frac{\sigma_1(z_1+1)}{\sigma_2 z_2}\right)^m\cdot\sum_{l=0}^{\infty}\left(\frac{\sigma_1 z_1}{\sigma_2(z_2+1)}\right)^l \tag{10.31}$$

The calculation is quite similar to the Gaussian case in (2.28) in Chap. 2.

The Gromov–Witten invariants of $\mathbf{P}^1$ for genus zero s-points have been studied up to $s = 4$ and they are [104]

$$
\begin{aligned}
\langle \tau_{2k_1} \tau_{2k_2} \rangle_{g=0} &= \frac{1}{k_1!^2 k_2!^2} \frac{1}{1 + k_1 + k_2} \\
\langle \tau_{2k_1-1} \tau_{2k_2-1} \rangle_{g=0} &= \frac{k_1 k_2}{k_1!^2 k_2!^2} \frac{1}{k_1 + k_2} \\
\langle \tau_{2k_1} \tau_{2k_2} \tau_{2k_3} \rangle_{g=0} &= \frac{1}{k_1!^2 k_2!^2 k_3!^2} \\
\langle \tau_{2k_1} \tau_{2k_2-1} \tau_{2k_3-1} \rangle_{g=0} &= \frac{k_2 k_3}{k_1!^2 k_2!^2 k_3!^2}
\end{aligned}
\tag{10.32}
$$

These results may be compared with the Gaussian means in (2.56), where τ_k is replaced by $\mathrm{tr} M^k$ with a normalization factor $k!$

References

1. M. Adler, P. Van Moerbeke, A matrix integral solution to two dimensional W_p-gravity. Commun. Math. Phys. **147**, 25–56 (1992)
2. M. Aganagic, R. Dijkgraaf, A. Klemm, M. Marino, C. Vafa, Topological strings and integrable hierarchies. Commun. Math. Phys. **261**, 451–516 (2006)
3. A. Alexandrov, Open intersection numbers, matrix models and MKP hierarchy, J. High Energy Phys. **03**, 042 (2015). arXiv:1410.1820
4. A. Alexandrov, Open intersection numbers, Kontsevich-Penner model and cut-and-join operators, J. High Energy Phys. **28** 25 (2015). arXiv:1412.3772
5. A. Alexandrov, Open intersection numbers and free field. arXiv:1606.06712
6. L. Álvarez-Gaumé, P. Basu, M. Marino, S.R. Wadia, Blackhole/string transition for the small Schwarzschild blackhole of $AdS_5 \times S^5$ and critical unitary matrix models. Eur. Phys. J. C **48**, 647 (2006)
7. J. Ambjørn, B. Durhuus, J. Fröhlich, Diseases of triangulated random surface models, and possible cures. Nucl. Phys. B **257**, 433–449 (1985)
8. J.E. Andersen, L.O. Chekhov, P. Norbury, R.C. Penner, Models of discretized moduli spaces, cohomological field theories, and Gaussian means, J. Geom. Phys. **98**, 312–339. Topological recursion for Gaussian means and cohomological field theories, Theoreticheskaya i Matematicheskaya Fizika **185**(3), 371–409 (2015)
9. A. Aptekarev, P. Bleher and A. Kuijlaars, Large n limit of Gaussian random matrices with external source, part II, Comm. Math. Phys. **259**, 367–389 (2005). arXiv:math-ph/0408041
10. J. Boussinesq, Théorie des ondes et des remous qui se propagent le long d'un canal rectangulaire horizontal, en communiquant au liquide contenu dans ce canal des vitesses sensiblement pareilles de la surface au fond. Jour. de Math. Pures et Appli. 55–108 (1872)
11. E. Brézin, *In Two Dimensional Quantum Gravity and Random Surfaces*, ed. By D.J. Gross, T. Piran and S. Weinberg (World Scientific, Singapore, 1992), p. 37
12. E. Brézin, D. Gross, The external field problem in the large-N limit of QCD. Nucl. Phys. B **97**, 120 (1980)
13. E. Brézin, C. Itzykson, G. Parisi, J.-B. Zuber, Planar diagrams. Commun. Math. Phys. **59**, 35–51 (1978)
14. E. Brézin, V.A. Kazakov, Exactly solvable field theories of closed strings. Phys. Lett. B **236**, 144–150 (1990)
15. E. Brézin, S.R. Wadia, The large N expansion in quantum field theory and statistical physics: from spin systems to 2-dimensional gravity, World Scientific Pub. Co. Inc. (1993)
16. E. Brézin, S. Hikami, A. Zee, Universal correlations for deterministic plus random hamiltonians. Phys. Rev. E **51**, 5442 (1995)
17. E. Brézin, S. Hikami, A. Zee, Oscillating density of states near zero energy for matrices made of blocks with possible application to the random flux problem. Nucl. Phys. **B464** [FS], 411 (1996)

E. Brézin and S. Hikami, *Random Matrix Theory with an External Source*,
SpringerBriefs in Mathematical Physics, DOI 10.1007/978-981-10-3316-2

18. E. Brézin, S. Hikami, Correlations of nearby levels induced by a random potential. Nucl. Phys. B **479**, 697–706 (1996)
19. E. Brézin, S. Hikami, Spectral form factor in a random matrix theory. Phys. Rev. E **55**, 4067 (1997). arXiv:cond-mat/9608116
20. E. Brézin and S. Hikami, Extension of level spacing universality, Phys. Rev. E **56**, 264 (1997). arXiv:cond-mat/9702213
21. E. Brézin, S. Hikami, Universal singularity at the closure of a gap in a random matrix theory. Phys. Rev. E **57**, 4140 (1998). arXiv:cond-mat/9804023
22. E. Brézin and S. Hikami, Level spacing of random matrices in an external source, Phys. Rev. E **58**, 7176 (1998). arXiv:cond-mat/9804024
23. E. Brézin, S. Hikami, A.I. Larkin, Level statistics inside the vortex of a superconductor and symplectic random matrix theory in an external source. Phys. Rev. B **60**, 3589 (1999). arXiv:cond-mat/9902037
24. E. Brézin, S. Hikami, Characteristic polynomials of random matrices. Commun. Math. Phys. **214**, 111 (2000). arXiv: math-ph/9910005
25. E. Brézin, S. Hikami, Characteristic polynomials of real symmetric random matrices. Commun. Math. Phys. **223**, 363 (2001). arXiv: math-ph/0103012
26. E. Brézin, S. Hikami, An extension of the Harish Chandra-Itzykson-Zuber integral. Commun. Math. Phys. **235**, 125–137 (2003)
27. E. Brézin, S. Hikami, New correlation functions for random matrices and integrals over supergroups. Commun. Math. Phys. **36**, 711 (2003). arXiv:math-ph/0208001
28. E. Brézin, S. Hikami, Vertices from replica in a random matrix theory. Commun. Math. Phys. **40**, 13545 (2007). arXiv:0704.2044
29. E. Brézin, S. Hikami, Intersection numbers of Riemann surfaces from Gaussian matrix models. Commun. Math. Phys. **10**, 096 (2007). arXiv:0709.3378
30. E. Brézin, S. Hikami, Intersection theory from duality and replica. Commun. Math. Phys. **283**, 507–521 (2008)
31. E. Brézin, S. Hikami, Intersection numbers from the antisymmmetric gaussian matrix model. Commun. Math. Phys. **07**, 050 (2008). arXiv:0804.4531
32. E. Brézin, S. Hikami, Computing topological invariants with one and two-matrix models. Commun. Math. Phys. **04**, 110 (2009). arXiv:0810.1085
33. E. Brézin, S. Hikami, Duality and replicas for a unitary matrix model. JHEP **07**, 067 (2010)
34. E. Brézin, S. Hikami, On an Airy matrix matrix model with a logarithmic potential. Commun. Math. Phys. **45**, 045203 (2012)
35. E. Brézin, S. Hikami, The Intersection numbers of the p-spin curves from random matrix theory. Commun. Math. Phys. **02**, 035 (2013)
36. E. Brézin, S. Hikami, Random matrix, singularities and open/close intersection numbers. Commun. Math. Phys. **48**, 475201 (2015)
37. R. Brower, P. Rossi, C.-I. Tan, The external field problem for QCD. Nucl. Phys. B **190**, 699 (1981)
38. A. Buryak, Equivalence of the open KdV and the open Virasoro equations for the moduli space of Riemann surfaces with boundary. Nucl. Phys. B **105**, 1427–1448 (2015). arXiv:1409.3888
39. A. Buryak, Open intersection numbers and wave function of the KdV hierarchy. Nucl. Phys. B **16**, 27–44 (2016). arXiv:1409.7957
40. J.B. Conrey, S.M. Gonek, Higher moments of the Riemann Zeta-function. Nucl. Phys. B **107**, 577–604 (2001)
41. A. D'Adda, A.C. Davis, P. Di Vecchia, P. Salomonson, An effective action for the supersymmetric CP^{n-1} model. Nucl. Phys. B **222**, 45–70 (1983)
42. F. David, Planar diagrams, two-dimensional lattice gravity and surface models. Nucl. Phys. **B257**(FS14), 45–58 (1985)
43. R. Dijkgraaf, Intersection theory, integrable hierarchies and topological field theory, new symmetry principles in quantum field theory. NATO ASI series, vol. 295 (Springer, Heidelberg, 1992). arXiv:hep-th/9201003

44. R. Dijkgraaf, H. Verlinde, E. Verlinde, Topological strings in d(1. Nucl. Phys. B **352**, 59–86 (1991)
45. R. Dijkgraaf, H. Verlinde, E. Verlinde, Loop equations and Virasoro constraints in non-perturbative two dimensional quantum gravity. Nucl. Phys. B **348**, 435 (1991)
46. R. Dijkgraaf, H. Verlinde, E. Verlinde, String propagation in a black hole geometry. Nucl. Phys. B **371**, 269–314 (1992)
47. M. Bertola, B. Dubrovin, D. Yang, Simple Lie algebras and topological ODEs. arXiv:1508.03750
48. M.R. Douglas, S.H. Shenker, Strings in less than one dimension. Nucl. Phys. B **335**, 635–654 (1990)
49. B. De Wit and G. 't Hooft, Nonconvergence of the 1/N expansions for SU(N) gauge fields on a lattice, Phys. Lett. B **69**, 61 (1977)
50. F.J. Dyson, A Class of Matrix Ensembles. J. Math. Phys. **13**, 90 (1972)
51. P. Desrosiers, B. Eynard, Super-matrix models, loop equations, and duality. J. Math. Phys. **51**, 123304 (2010). arXiv:0911.1762
52. P. Desrosiers, Duality in random matrix ensembles for all β, Nucl. Phys. B **817**, 224 (2009). arXiv:0801.3438
53. X-M. Ding, Y. Li and L. Meng, From r-spin intersection numbers to Hodge integrals, JHEP 01(2016)015. arXiv:1507.04093
54. L.J. Dixson, J. Lykken, M.E. Peskin, N = 2 superconformal symmetry and SO(2,1) current algebra. Nucl. Phys. B **325**, 329 (1989)
55. M. Douglas, Strings in less than one dimension and the generalized KdV hierarchies. Nucl. Phys. B **238B**, 176 (1990)
56. V.G. Drinfeld, V.V. Sokolov, Lie algebras and equations of Korteweg-de Vries type. Nucl. Phys. B **30**, 1975–2036 (1985)
57. T. Eguchi Hori, S-K. Yang, Topological σ models and large-N matrix integrals, arXiv:hep-th/9503017
58. C. Faber, S. Shadrin, D. Zvonkine, Tautological relations and the r-spin Witten conjecture. Nucl. Phys. B **43**, 621 (2010)
59. H. Fan, T. Jarvis and Y. Ruan, The Witten equation, mirror symmetry and quantum singularity theory, Annals of Math. 178 (2013),1-106. arXiv:0712.4021 [math.AG]
60. V.A. Fateev, A.B. Zamolodchikov, Conformal quantum field theory models in two-dimensions having Z(3) symmetry. Nucl. Phys. B **280**, 644 (1987)
61. P.J. Forrester, *Log-Gases and Random Matrices, (LMS-34) London Mathematical Monographs* (Princeton University Press, Princeton, 2010)
62. M. Fukuma, H. Kawai, R. Nakayama, Infinite dimensional Grassmannian structure of two dimensional quantum gravity. Commun. Math. Phys. **143**, 371 (1992)
63. D. Gaiotto, L. Rastelli, A paradigm of open/closed duality Liouville D-branes and the Kontsevich model. JHEP **07**, 053 (2005). arXiv:hepth/0312196
64. I.M. Gel'fand, L.A. Dikii, Asymptotic behavior of the resolvent of Strum-Liouville equations and the algebra of the Korteweg-De Vries equations. Russ. Math. Surv. **30**:5, 77 (1975)
65. I.M. Gel'fand and L.A.Dikii, Fractional powers of operators and hamiltonian systems, Funktsional'nyi Analiz i Ego Prilozheniya **10**(4), 13 (1976)
66. I.M. Gel'fand, L.A. Dikii, The resolvent and Hamiltonian systems. Funktsional'nyi Analiz i Ego Prilozheniya **112**, 11 (1977)
67. A. Givental, A_{n-1} singularities and n KdV hierarchies. Mosc. Math. J. **32**, 475–505 (2003). arXiv:math.AG/0209205
68. I.P. Goulden, J.L. Harer, D.M. Jackson, A geometric parametrization for the virtual Euler characteristics of the moduli spaces of real and complex algebraic curves. Commun. Math. Phys. **353**, 4405–4427 (2001). arXiv:math/9902044
69. D.J. Gross, A.A. Migdal, Nonperturbative two dimensional quantum gravity. Phys. Rev. Lett. **64**, 127–130 (1990)
70. D.J. Gross, M.J. Newman, Unitary and Hermitian matrices in an external field. 2: the Kontsevich model and continuum Virasoro constraints. Nucl. Phys. B **380**, 168 (1992). hep-th/9112069

71. D.J. Gross, E. Witten, Possible third order phase transition in the large-N lattice gauge theory. Nucl. Phys. B **21**, 446 (1980)
72. M.A. Halasz, J.J.M. Verbaarschot, Effective Lagrangians and chiral random matrix theory physics. Nucl. Phys. B **52**, 2563 (1995)
73. A. Hanany, N. Prezas, J. Troost, The partition function of the two dimensional black hole conformal field theory. Nucl. Phys. B **04**, 014 (2002)
74. G.H. Hardy, On certain deffinite integrals considered by Airy and by Stokes. Nucl. Phys. B **41**, 226–240 (1910)
75. G. Harder, A Gauss-Bonnet formula for discrete arithmetically defined groups. Nucl. Phys. B **4**, 409–455 (1971)
76. J. Harer, D. Zagier, The Euler characteristics of the moduli space of curves. Nucl. Phys. B **85**, 457–485 (1986)
77. H. Chandra, Invariant differential operators on a semisimple lie algebra. Proc. Nat. Acad. Sci. **42**, 252–253 (1956)
78. S. Hikami, E. Brézin, WKB-Expansion of the Harish Chandra - Itzykson-Zuber integral for arbitrary β. Nucl. Phys. B **116**, 441 (2006)
79. Y. Ikhlef, J.L. Jacobsen, H. Saleur, An integrable spin chain for the SL(2, R)/U(1) black hole sigma model. Phys. Rev. Lett. **108**, 081601 (2012). arXiv:1109.1119
80. C. Itzykson, J.-B. Zuber, The planar approximation II. J. Math. Phys. **21**, 411–421 (1980)
81. H. Jack, A class of symmetric polynomial with a parameter. J. Math. Phys. **69**, 1–18 (1970)
82. D. Jackson, T.I. Visentin, *An Atlas of the Smaller Maps in Orientable and Nonorientable Surfaces* (Chapman and Hall/CRC, Boca Raton, 2001)
83. M. Jimbo, T. Miwa, Y. Mori, M. Sato, Density matrix of an impenetrable Bose gas and the fifth Painlevé transcendent. Phys. D Nonlinear Phenom. **1**, 80 (1980)
84. V.A. Kazakov, A.A. Migdal, Recent progress in the theory of noncritical strings. Nucl. Phys. B **3111**, 171–190 (1988/1989)
85. J.P. Keating, N.C. Snaith, Random matrix theory and $\zeta(1/2+it)$. Commun. Math. Phys. **214**, 57–89 (2000)
86. T. Kimura, X. Liu, A genus-3 topological recursion relation. Commun. Math. Phys. **262**, 645 (2006)
87. T. Kimura, Note on a duality of topological branes, PTEP (2014) 103B04, arXiv:1401.0956 [hep-th]; Duality and integrability of a supermatrix model with an external source, PTEP (2014) 123A01, arXiv:1410.0680 [math-ph]
88. V.G. Knitzhnik, A.B. Zamolodchikov, Current algebra and Wess-Zumino model in two dimensions. Nucl. Phys. B **247**, 83 (1984)
89. M. Kontsevich, Intersection theory on the moduli space of curves and the matrix Airy function. Commun. Math. Phys. **147**, 1–23 (1992)
90. I.K. Kostov, Matrix models as conformal field theories, in *Applications of Random Matrices in Physics*, ed. By E. Brezin et al. *Proceedings of the NATO Advanced Study Institute on Application of Random Matrices in Physics*, Les Houches, France (Springer, 2006)
91. K. Liu, R. Vakil, H. Xu, From pseudo differential operators and Witten's r-spin numbers. J. fRr die reine und angewandte Mathematik, Crelle-2014-0102. arXiv:112.4601
92. K. Liu, H. Xu, Descendent integrals and tautological rings of moduli spaces of curves. Geom. Anal. Adv. Lect. Math. (ALM) **18**2, 137–172 (2010). arXiv:0912.0584
93. I.G. Macdonald, *Symmetric Functions and Hall Polynomials* (Oxford University Press Inc., New York, 1995)
94. J. Maldacena, H. Ooguri, J. Son, Strings in AdS_3 and the SL(2, R) WZW model. II: euclidean black hole. J. Math. Phys. **42**, 2961 (2001)
95. J. Maldacena, G. Moore, N. Seiberg, D. Shih, Exact versus semiclassical target space of the minimal string. JHEP **10** (2004)
96. M. Marino, Chern-Simons Theory, Matrix Models, and Topological Strings. Oxford University Press (2005)
97. M.L. Mehta, *Random Matrices*, 3rd edn. (Academic Press, Elsevier, New York, 2004)

98. A. Mironov, A. Morozov, G.W. Semenoff, Unitary matrix integrals in the framework of the generalized Kontsevich model. 1: Brezin-Gross-Witten model. Int. J. Mod. Phys. A **11**, 5031 (1996)
99. M. Mirzakhani, Weil-Petersson volumes and intersection theory on the moduli space of curves. Int. J. Mod. Phys. A **20**, 1–23 (2007)
100. H.L. Montgomery, The pair correlation of zeros of the zeta function, Analytic number theory, St Louis University, 1972 (ed. H.G. Diamond), Proceedings of Symposia in Pure Mathematics 24 (American Mathematical Society, Providence, RI, 1973) 181-193
101. A. Morozov, Integrability and matrix models. Phys. Usp. **37**, 1–55 (1994). hep-th/9303139
102. A. Morozov, S. Shakirov, Exact 2-point function in Hermitian matrix model. JHEP **12**, 003 (2009). arXiv:0906.0036
103. A. Morozov, S. Shakirov, From Brezin-Hikami to Harer-Zagier formulas for Gaussian correlators. arXiv:1007.4100
104. P.N. Norbury, N. Scott, Gromov-Witten invariants of P^1 and Eynard-Orantin invariants. Geometry and Topology **18**, 1865–1910 (2014). arXiv: 1106.1337
105. A.M. Odlyzko, On the distribution of spacings between zeros of the zeta function. Phys. Usp. **48**, 273–308 (1987)
106. A. Okounkov, Random matrices and random permutations. Phys. Usp. **20**, 1043–1095 (2000)
107. A. Okounkov, Generating functions for the intersection numbers on moduli spaces of curves. Phys. Usp. **18**, 933–957 (2002)
108. A. Okounkov and R. Pandharipande, Gromov-Witten theory, Hurwitz numbers, and Matrix models I, Proceedings of symposia in pure mathematics Vol.80 Algebraic Geometry Seattle 2005, American Mathematical Society. math.AG/0101147
109. A. Okounkov, R. Pandharipande, Gromov-Witten theory, Hurwitz theory and completed cycles. Ann. Math. **163**, 517 (2006). AG/0204305
110. A. Okunkov, N. Reshetikhin, Random skew plane partitions and the Pearcey process. Phys. Usp. **269**, 571–609 (2007). arXiv:math/0503508
111. The Oxford Handbook of Random Matrix Theory, edited by G.J. Baik, P. Di Francesco. (Oxford University Press, Akemann, 2011)
112. R. Pandharipande, J.P. Solomon, R.J. Tessler, Intersection theory on moduli of desks, open KdV and Virasoro. arXiv:1409.2191
113. L.A. Pastur, On the spectrum of random matrices. Theor. Math. Phys. **10**, 67–74 (1972)
114. R.C. Penner, Perturbative series and the moduli space of Riemann surfaces. Theor. Math. Phys. **27**, 35 (1988)
115. T. Pearcey, The structure of an electromagnetic field in the neighbor hood of a cusp of a caustic. Theor. Math. Phys. **37**, 311–317 (1946)
116. A.M. Polyakov, P.B. Wiegmann, Goldstone fields in two dimensions with multivalued actions. Theor. Math. Phys. **141B**, 223 (1984)
117. P.M. Bleher, A.R. Its (ed.), Random Matrix Models and Their Application, vol. 40 (MSRI publications, Cambridge University Press, Cambridge, 2001)
118. B. Riemann, Bernard Riemann's Gesammelte Mathematische Werke und Wissenschaftlicher Nachlass (Cambridge Library Collection-Mathematics)
119. B. Safnuk, Topological recursion for open intersection numbers. arXiv:1601.04049
120. S. Samuel, U(N) Integrals, 1/N, and the De Wit-'t Hooft anomalies. J. Mathe. Phys. **21**, 2695 (1980)
121. I. Satake, The Gauss-Bonnet theorem for V-manifolds. J. Math. Phys. **9**, 464–492 (1957)
122. C.L. Siegel, Symplectic Geometry. Am. J. Math. 65, 1–86 (1943). Gesammelte Werke, Bd. III, s. 443–458
123. L. Susskind, *Matrix theory black holes and the Gross Witten transition, in Nonperturbative aspects of strings, branes and supersymmetry, Trieste 1998* (World Scientific, Singapore, 1998). hep-th/9805115
124. T. Tao, *Topics in Random Matrix Theory*. Graduate Studies in Mathematics vol. 132 (American Mathematical Society, Providence, 2012)
125. G. t' Hooft, A planar diagram theory for strong interactions. Nucl. Phys. B **18**, 461–473 (1974)

126. C.A. Tracy, H. Widom, Level-spacing distributions and Airy Kernel. Commun. Math. Phys. **159**, 151 (1994)
127. C.A. Tracy, H. Widom, The pearcey process. Commun. Math. Phys. **263**, 381–400 (2006)
128. G. Van der Geer, Cycles on the moduli space of Abelian varieties, Moduli of curves and abelian varities, Vol. 33 of the series Aspects of Mathematics, 65–89 (1999). arXiv:alg-geom/9605011
129. G.N. Watson, *Theory of Bessel Function* (Cambridge University Press, Cambridge, 1922)
130. F. Wegner, *Supermathematics and Its Applications in Statistical Physics*. Lecture Notes in Physics, vol. 920 (Springer, Berlin, 2016)
131. J. Wess, B. Zumino, Consequences of anomalous ward identities. Phys. Lett. B **37**, 95 (1971)
132. E.P. Wigner, On the Statistical distribution of the widths and spacings of nuclear resonance levels. Phys. Lett. B **47**, 790 (1951)
133. E. Witten, Non-abelian bosonization in two dimensions. Commun. Math. Phys. **92**, 455 (1984)
134. E. Witten, Two dimensional gravity and intersection theory on moduli space. Commun. Math. Phys. **1**, 243–310 (1991)
135. E. Witten, Algebraic geometry associated with matrix models of two dimensional gravity, in *Topological Methods in Modern Mathematics* (Publish or Perish, INC., 1993), pp. 235–269
136. E. Witten, The N matrix Model and Gauged WZW Models. Nucl. Phys. B **371**, 191 (1992)
137. E. Witten, String theory and black holes. Nucl. Phys. B **44**, 314 (1991)
138. A.B. Zamolodchikov, V.A. Fateev, Nonlocal(parafermion) currents in two-dimensional conformal quantum field theory and self-dual critical points in Z_N-symmetric statistical systems. Nucl. Phys. B **62**, 215 (1985)
139. J. Zhou, Solution of W-constraints for r-spin intersection numbers. arXiv:1305.6991

Index

E. Brézin and S. Hikami, *Random Matrix Theory with an External Source*,
SpringerBriefs in Mathematical Physics, DOI 10.1007/978-981-10-3316-2

Printed in the United States
By Bookmasters